WITHDRAWN FROM

KT-473-919

KA 0291901 X

$ELLING $EATTLE

$ELLING $EATTLE

REPRESENTING CONTEMPORARY URBAN AMERICA

James Lyons

 **WALLFLOWER PRESS** LONDON & NEW YORK

First published in Great Britain in 2004 by
Wallflower Press
4th Floor, 26 Shacklewell Lane, London E8 2EZ
www.wallflowerpress.co.uk

Copyright © James Lyons 2004

The moral right of James Lyons to be identified as the author of this work has been asserted in
accordance with the Copyright, Designs and Patents Act of 1988

All rights reserved. No part of this publication may be reproduced, stored in a retrieval system,
or transported in any form or by any means, electronic, mechanical, photocopying, recording or
otherwise, without the prior permission of both the copyright owner and the above publisher of
this book.

A catalogue for this book is available from the British Library

ISBN 1-903354-96-5

Book design by Elsa Mathern

Printed by Edwards Brothers Inc., Ann Arbor, Michigan, USA

UNIVERSITY COLLEGE
WINCHESTER

9797
LYO   029190 X

CONTENTS

ACKNOWLEDGEMENTS

I am indebted to a number of people who contributed in helping this book to completion. For their guidance and support from the very early stages of research, I thank Mark Jancovich and Douglas Tallack at the University of Nottingham. I also thank the Arts and Humanities Research Board for their financial assistance, without which I would not have been able to undertake the research project that developed into this book. I have benefited immeasurably from the attentive readings and insightful commentary of a number of individuals. Paul Grainge and Martyn Bone offered typically helpful and perceptive suggestions along the way, as well as friendship of the highest order. Roberta Pearson's invaluable proposals for improvement are very much appreciated. I would also like to thank several anonymous readers for providing numerous astute suggestions for revision. Yoram Allon and Hannah Patterson at Wallflower Press have been tremendously enthusiastic and encouraging.

My thanks also extend to Paul Jaye for his warm hospitality during my stay in Seattle, and for helping me to see (yet) another side to the city, defined liberally as 'ephemeral research'. I would like to pay special thanks to my family for their love and support, and to the Goldsteins and the Avigads for all their kindness and munificence. Finally, I dedicate this book to Karen Goldstein, to whom I owe more than words can say.

INTRODUCTION

Doug: It's time to go.

Carol: Where?

Doug: I liked Seattle when I was in Med. School. I didn't see much of it, and I liked it.

Carol: Seattle!?

Doug: The Pacific Northwest … It's pretty up there … it's clear … you'd like it. Come with me.

– *E.R.*, 'The Storm'[1]

Removing a main character from a successful television show can be a difficult theatrical procedure, none more so than the complex operation required to extricate Doug Ross (George Clooney) from the Chicago-set prime-time hospital drama *E.R.* in 1998. The excision was fraught and arduous, yet, interestingly, the writers sought to suggest that solace was on hand in the form of knowledge that Ross would go on to 'a better place', namely Seattle. It would be hard to imagine a similar show making a corresponding reference to Seattle even ten years earlier. That *E.R.* could fashion such a passing mention of the city, and be assured of its audience understanding what leaving for Seattle 'meant', was a decidedly contemporary phenomenon. The 1990s saw the city establish a new prominence within American popular culture: the location for a succession of headline grabbing events; the site for the genesis of new trends and developments; the subject of an increased number of magazine profiles; and the signified setting for a notable number of novels, motion pictures, television shows and documentaries. In selecting Seattle from a wide range of potential locations, *E.R.* thus revealed much about the city's status and reputation in the 1990s. Painful as it was, when Ross stated he was leaving for Seattle, *E.R.*'s audience understood.

When *E.R.*'s writers opted for Seattle as Doug Ross's haven from the trials and tribulations of Chicago, it is clear that they were drawing heavily on the contrast the city would be guaranteed to provide with its Midwestern counterpart. Over the preceding decade Seattle had gained a powerfully mythic reputation as an eminently 'liveable' and desirable city at the forefront of contemporary culture. As the locale for a profusion of new technology companies, the most prominent being the software behemoth Microsoft, Seattle appeared to be a model for the New American Economy, a hub for the sort of innovation, entrepreneurial spirit and élan that would lead the nation forward into the intensified competition for jobs, markets and profits in the global marketplace. Moreover, in the guise of its phenomenally successful coffee retailer Starbucks, the city excelled in branding and exporting a particular vision of stylish urban living to cities across the US, and increasingly throughout Western Europe and Asia. This was aided and abetted by the city's fictional representation in NBC's Emmy award-winning prime-time sitcom *Frasier*, which made much of Seattle's growing reputation as the epicentre of 'coffee culture'. High-profile, high-grossing Hollywood films such as *Sleepless in Seattle* (1993) and *Disclosure* (1994) also served to consolidate a sense of the city as an enviable site for upscale living. Seemingly at the other end of the spectrum, the phenomenal success of so-called Grunge music, represented by bands such as Pearl Jam, Soundgarden and, above all, Nirvana, nevertheless worked to focus media attention on the city as a vibrant, trend-setting urban locale.

Seattle's high profile and status in the 1990s should be set in the context of the fluctuating fortunes of the city throughout its history – a history characterised by successive waves of boom and bust, with Seattle always seemingly on the cusp of national and international prominence. Founded by settlers in 1851, Seattle first became a major commercial centre by way of its strategic position for those trekking north for the Yukon Gold Rush in 1897. By 1909 the city held the Alaska-Yukon-Pacific Exposition, a civic fair that drew almost four million visitors, and was, according to local historian Roger Sale, intended to "proclaim to the world that Seattle had arrived".[2] Yet the desire on the part of the exposition's organisers to engineer a Seattle with "an unending bright future" was not realised.[3] As Sale notes, less than ten years later "Seattle had reached a dead end. Its business and manufacture reverted to an emphasis on basic extractive industry … and its growth was at a standstill."[4]

Economic hardship in the aftermath of World War One combined with strong labour unions and a populace sympathetic to radical politics to produce Seattle's next brush with fame when in February 1919 the city became the first in the nation to declare a general strike, eventually lasting five days. When the strike ended, Seattle was left with economic stagnation that continued until World War Two, when defence spending and federal contracts stimulated urban growth and development along the west coast of the US.[5] In Seattle, that meant Boeing, who gained war contracts to roll out thousands of war planes from its factories in Seattle, Everett and Renton. Indeed, throughout the post-war period the city's prosperity was tied closely to Boeing, who dominated the local economy, and bestowed Seattle with the sobriquet 'Jet City'. This status also served as impetus for the decision to hold the 1962 World's Fair in Seattle, yet again "proposed as an economic stimulant and a key step in Seattle's rise to metropolitan big-league status … [and] to help Seattle overcome a reputation as provincial and unsophisticated".[6] The World's Fair was something of a mixed success in attaining this goal, bequeathing the city with new icons of urban legibility, such as the Monorail and the Space Needle (a futuristic tower with a revolving restaurant at the top), and a temporarily heightened profile. However, it did not fulfil the objective of establishing a broader and more diversified local economy. Indeed, by the late 1960s and early 1970s, the widespread economic recession that poleaxed many of the nation's urban centres was in Seattle called the 'Boeing depression', and which hit the city so hard that the area began to witness sightings of the bumper stickers and billboards (now long passed into local lore) that "asked the last person leaving Seattle please to turn out the lights".[7]

The history of Seattle's depiction in popular culture also reveals a city that suffered for a long time with an image as a parochial and insular place. For instance, the 1937 movie *Stage Door* saw Lucille Ball playing Judy Canfield, a fledgling actress struggling to succeed in Hollywood. Interpreting her lack of professional success as prejudice against her city of origin, Judy exclaims at one point "Am I supposed to apologise for being born in Seattle?" to which her friend Eve (Eve Arden) replies "I thought the people out there lived in trees."[8] When renowned British conductor Sir Thomas Beecham was brought to Seattle to lead the city's symphony orchestra in the early 1940s, he quickly resigned, but not before offering an infamous farewell, declaring the city an "aesthetic dustbin".[9] In J. D. Salinger's celebrated 1951 novel *The Catcher in the Rye*, Holden Caulfield encounters three young female tourists in New York's 5 Club. Portrayed as both naïve and dull, Caulfield finds out that they are from Seattle. His parting aside is to state that "I told them I'd look them up in Seattle sometime, if I ever got there, but I doubt if I ever will".[10] Caulfield was not alone in his decision to give Seattle a wide berth: between the release of MGM's *Tug Boat Annie* in 1933 and the Elvis Presley vehicle *It Happened At the World's Fair* in 1962, there were no major motion pictures set in or shot on location in Seattle.[11] Moreover, as recently as 1984, Brian De Palma's thriller *Body Double* depicted a character exiting Los Angeles for Seattle being told to go and "knock 'em dead", to which he answered with the riposte, "they're *already* dead in Seattle" – a comment that served to reinforce a perception of the city as a decidedly lacklustre backwater.[12]

Yet by 1996 *Newsweek* had placed Seattle on its front cover, with the headline 'Swimming to Seattle' and posed the question: "Everybody Else is Moving There. Should You?"[13] Clearly, something had changed in the intervening period. In this book I examine the nature of that change, in the process tracing the trajectory to Seattle's assured new presence in American culture in the 1990s. My interest in Seattle began simply with an intuitive sense that the city *had* become a more high-profile location over recent times, which provoked a desire to determine whether this was borne out by evidence of an upsurge in exposure. What soon became apparent was not only that the excess of material attested to the proliferation of films, novels, documentaries and news profiles depicting the city, but that an attempt to account for and make sense of the entire range of Seattle's media coverage over this period would be an unfeasible task. However, what also became evident from the perusal of a number of the more popular, critically acclaimed, and influential representations of Seattle in the 1990s was that they seemed to have one important thing in common. This was the way in which they worked to connect Seattle in interesting and original ways to wider debates concerned with the current state and future shape of America's large urban centres. In other words, what seemed to be clear was that Seattle's rise to fame was in a fundamental sense tied to more widespread anxieties about the fate of the contemporary American city.

While general concern about the condition and future direction of American cities is nothing new, it is also the case that the last few decades have witnessed profound changes in the nation's major urban centres.[14] The 1980s and 1990s saw widescale and often painful economic and social restructuring radically re-organise the country's metropolitan areas. Succinctly captured by James Donald's phrase 'the three Gs', the forces of globalisation, gentrification and ghettoisation served to transform cities and their populations in dramatic fashion.[15] Municipal governments worked feverously to promote their cities as headquarters for corporations engaged in global commerce and financial services, attuned to the new logics of information and transportation technologies and the advanced flexibility of labour and markets. Yet commentators also noted the increasing extremes of wealth and poverty that characterised contemporary cities, exacerbating the already polarised nature of the urban fabric. The agglomeration of high-tech, high-revenue-generating corporate control centres in US cities was linked to the establishment of new housing and leisure enclaves; waterfront developments, festival 'marketplaces' and upscale apartment complexes – gentrified spaces of consumption and privatised areas of urban development enjoyed by the fortunate few.[16] These developments were in turn related to the creation of a permanent underclass composed in part by an influx of first-generation immigrants from South-east Asia and Latin America, working to support these new functions in a range of low-paying service occupations, segregated into new urban ghettoes, and largely excluded from these privatised public spaces and lifestyle enclaves. Moreover, this widening economic and social divide was occurring against the backdrop of increased anxieties over deteriorating infrastructure, environmental pollution, rising crime and gang violence, drug addiction, escalating homelessness and the inexorable flight of the white middle class from the central cities.

The early 1990s arguably represents something of a nadir in terms of the confluence of trends and developments which were seen to represent crisis and failure in urban America – nowhere more apparent than in the two cities that have tended to dominate academic and popular commentaries on the state of contemporary urban America, namely Los Angeles and New York. New York lurched through a severe financial crisis in the early 1990s, losing over 250,000 jobs between 1991 and 1992, a figure unprecedented for an American city. In addition, publicity over the city's spiralling crime rate, crack addiction figures and AIDS infection rates helped hasten the demise of the scandal-hit Democrat administration led by Mayor David Dinkins and paved the way for the election of the Republican Rudolph Giuliani and his much publicised 'zero-tolerance' policing policy. Los Angeles found its own dramatic crisis point in the serious civil unrest that accompanied the 'not guilty' verdict for the police officers indicted in the Rodney King trial of 1992. The decision by an all-white jury to acquit four white officers caught on camera assaulting a black motorist became the catalyst for the largest instance of civil unrest since the Watts riots in 1965, and brought to a head the long-simmering antagonism between various sections of the populace at the sharp-end of the unevenly developed metropolis and those charged with maintaining what Mike Davis has famously termed the city's 'spatial apartheid' of rich and poor.[17] Liam Kennedy, who has looked closely at how both cities were represented on film and in literature at this time, writes that during the period New York and Los Angeles "began to appear in cultural representations as emblematic sites of strain and fracture in the symbolic order of the national culture".[18] Davis, noting the propensity of Hollywood to indulge in the on-screen annihilation of Los Angeles by flood, fire or earthquake during this period, observes that "no city, in fiction or film, has been more likely to figure as the icon of a really bad future (or present, for that matter)", seeing in the literary and cinematic destruction of Los Angeles an anxiety "that leads us back to the real Los Angeles as well as to the deepest animating fears of our culture".[19] Such depictions drew upon the material conditions in these major cities in intricate ways and worked narratives and metaphors of urban crisis into representational form, in the process mediating the complex relationship between the city as material environment and as imaginary space.[20]

In an important sense this period saw Los Angeles and New York function as focal points for what Robert Beauregard has termed a "discourse on urban decline" influential in shaping popular attitudes towards problems in the nation's large cities over the last few decades. Beauregard argues that during this period this discourse functioned "ideologically to shape understandings of developments in cities" and to reveal their role as "focal points for the nation's collective anxieties".[21] Indeed, he maintains that even in the mid-1980s, as gleaming new office blocks and apartment complexes sprung up in cities, and "the striking affluence and new urban landscapes" seemed momentarily to have displaced notions of urban deterioration, "persistent and deepening poverty, the solidification of a permanent under-class, the continued existence of slums, the rise of new immigrant neighbourhoods, [and] the unrelenting decline of the white population" maintained a considerable and powerful contrast, and served only to reinforce a sense of the widening schisms in the urban fabric.[22]

He notes that by the late 1980s "drugs, crime, racial tension and poverty" were again the defining characteristics of the major cities, and by the early 1990s, with a new economic recession placing cities such as New York and Philadelphia on the edge of bankruptcy and the riots in Los Angeles bringing race urgently to the forefront of concerns over urban decline, public debates over the nation's large urban centres were thus characterised by a "return to disbelief, indecision and waning hope".[23]

It is no coincidence that the period in which New York and Los Angeles were perceived to be most dramatically and spectacularly in crisis was also that which represented Seattle's most intensive phase of coverage and exposure in the American press – this is also the moment when many of the most influential and successful films, novels and television shows examined in this book appeared. Such representations drew selectively, and with varying degrees of fantasy and fidelity, upon changes and developments in the city, and worked to shape the various meanings of Seattle available to national and international audiences, readers and consumers. Crucially it is in this period that Seattle's role as an organising site and symbolic repository for an enduring set of fears, hopes and desires relating to the changing form and function of America's urban centres came sharply into view. Emerging to prominence during this time of upheaval, Seattle appeared within popular accounts as a city seemingly largely unscathed by the negative repercussions of the profound economic and social restructuring affecting other American cities, serving as an alternative vision of the nation's urban present and its potential urban future.

It is important to recognise that Seattle was by no means the only city to be identified as a counterpoint to prevailing trends in urban restructuring at this time. The period saw a number of 'middle tier' cities, many located in the south and west, cited repeatedly in the US media for noteworthy trajectories of economic growth and coveted 'quality of life' indicators. 'Smaller faster business cities' such as Atlanta, Dallas, Denver, Houston, Las Vegas and San José were marked out for attention due to their apparent ability to capitalise on growth in service-sector employment and developments in the field of advanced technologies.[24] In addition, Seattle's neighbours to the south and north, namely Portland and Vancouver, both received praise for their combination of perspicacious urban planning, laudable amenities and enviable economic buoyancy. Indeed, all three found themselves united as the focal point for commentators' projections of an emerging Pacific Northwest city-region: a "trans-border economic and political powerhouse" anticipated to materialise in decades to come.[25] More prosaically, Vancouver, in matching Seattle's extolled landscaping and scenery, and surpassing it for well-maintained infrastructure (arguably as a result of the greater emphasis placed on public programmes within Canadian municipal culture), served occasionally to remind the US media of the limits of its own symbol of national urban revival.[26] Yet none of these cities could rival Seattle for prominence or profile, and for the way that it seemingly excelled in fusing economic vitality with modish cachét. For example, it is worth noting that September 1993 saw the first broadcast of two of the most popular television shows of the 1990s, namely *Frasier* (1993–2004) and *The X-Files* (1993–2002). While *Frasier*'s Seattle setting drew heavily on the city's voguish appeal, *The X-Files* was filmed in Vancouver (for lower production costs) but its setting remained resolutely in California. More tellingly, *The X-Files*' creator Chris Carter's next network show, entitled *Millennium* (1996–1999), would again be shot in Vancouver, but this time would be set in Seattle. An isolated example, perhaps, but it nevertheless serves to illustrate the fact that Seattle clearly possessed a representational currency (within the US media, it should be said) that its northern neighbour could not emulate.

In this book I examine many of the most popular, significant and critically acclaimed depictions of Seattle in the 1990s, and consider some of the most important trends and developments that were crucial to making Seattle a more visible and more visibly represented city during the period. The majority of the films, novels, magazines and television shows I discuss could be regarded as 'mainstream', including nationally and internationally distributed general current affairs and news publications such as *Time* magazine, *Newsweek* and the *New York Times*; prime-time major network television series such as *Frasier* and *Millennium*; and major motion pictures such as *Sleepless in Seattle* (1993), *Disclosure* (1994) and *The Hand that Rocks the Cradle* (1992). 'Mainstream' is a term that requires a cautious and judicious application, and is used here to signal that the book's central focus is not upon 'alternative', 'underground', 'micro' or 'radical' media, all of which often explicitly position themselves in opposition to existing local or national media forms. This is an important caveat for the present study: during the 1990s Seattle, like a number of American urban centres, became a hub for a range of 'alternative' media networks and forms of cultural production. Elsewhere I have discussed the ways in which such underground media production functioned within the micro-politics of neighbourhood cinema exhibition sites – in particular, the ways in which digital cinema production was taken up within radical political discourse.[27] In contrast, one of my main concerns in this book is with charting the contours of what could be termed a mainstream media discourse on Seattle.[28] What I am particularly interested in here is the ways in which the city gained an unprecedented level of exposure within the primary channels of news and entertainment distribution in the US in the 1990s, and to question precisely the nature of its new found appeal. However, this is not to suggest that the book seeks merely to chart the smooth progress of an unproblematic consensus concerning the city; on the contrary, it is attentive to the ways in which the selected texts examined here often suggested significant moments of contradiction and contestation, and worked to problematise prevailing understandings of Seattle's relationship to key urban trends and transformations.

The book is divided into six chapters, each of which approaches Seattle from the perspective of an important theme or topic that can be seen to unite a range of significant depictions of the city. The first four chapters address themes that clearly convey Seattle's relationship to wider debates and issues relating to the contemporary city. For instance, chapter one's concern with Seattle's portrayal as a high-tech city on the cutting edge of contemporary culture positions it in relation to widespread anxieties over the impact of new technologies on city development and growth, while chapter two's focus on Seattle's avowed relationship to its beautiful natural surroundings situates it in the context of 1990s concerns with urban pollution and environmental degradation, as well as significant trends in nature tourism and recreation. Chapter three's concern with the question of Seattle's racial profile, and chapter four's look at the city's reputation as an idyll for family life locates it similarly with reference to key questions regarding the transformed demographic composition of America's urban centres over the last twenty years. The last two chapters, which focus in turn on grunge music and the emergence of the city's reputation as the centre of a new American 'coffee culture', are ostensibly more specifically Seattle in origin, but what underpins both of these important cultural phenomena are their links to wider shifts in urban economies geared increasingly towards new consumption practices and the branding of 'unique' place identities. Thus while the connecting thread between all the chapters is the way in which the themes and topics addressed reveal crucial perspectives on Seattle's rise to fame in the 1990s, what makes them of real import is the fact that they demonstrate the city's function as the locus for a precise range of responses to the perceived crises of American urbanism in the 1980s and 1990s. In this way the book makes a contribution towards the growing body of literature that considers the role played by cities other than New York, Los Angeles and Chicago in the formation of shared and contested meanings of contemporary urbanism, a body of work which has so far left Seattle largely unexamined. It is also able to present the evidence for the need to include Seattle within this body of work, not only for the way in which it highlights a quite different range of responses to the challenges posed by profound socio-economic restructuring to those found by examining cities such as New York, Los Angeles and Chicago, but also for the ways in which it is thus able to engage with and contribute to a number of important debates within the field of urban studies.

The acclaimed British writer Jonathan Raban, who moved to Seattle in the late 1980s, has suggested that the city and the Pacific Northwest region of the US used to be "a blank in people's imaginative maps". Speaking in 1992, Raban noted that "recently [it is] gaining a kind of visual identity. Bit by bit, this land is being made public and mythical, like New England of the nineteenth century, or the American South".[29] What I seek to show in this book is that this process of mythification did not simply suggest the city's mysterious capacity to embody the 'zeitgeist' in the 1990s, but rather underscored the relationship between Seattle, the various media forms which sought to represent the city, and the complex and often contradictory range of discourses at play in the wider culture during the period. If *Newsweek*'s claim that "everybody's moving to Seattle" was journalistic hyperbole, it was perhaps nearer the truth to suggest that in the 1990s Seattle seemed to be moving closer to everybody.

CHAPTER 1

ON THE EDGE WITH A FEAR OF FALLING:
SEATTLE AND NEW TECHNOLOGY

In February 2000 Microsoft held a launch party for the latest version of its Windows operating system, named, appropriately enough, Windows 2000. The largest and most powerful software company in the world, the Seattle-based Microsoft corporation's major software launches were events relayed in newspapers and periodicals across the world. At the party in San Francisco, Bill Gates, Microsoft's co-founder and CEO, stood in front of a large digital screen, on to which the dazzling graphics demonstrating the 'look and feel' of Windows 2000 were unveiled. Privileged among the range of digitally generated illustrations was a sequence that depicted Seattle's Space Needle against the glittering backdrop of the night-time cityscape. The Space Needle, a slender, iconic tower constructed for Seattle's Century 21 Exposition in 1962, (also known as the Seattle World's Fair) was the centrepiece of Microsoft's shimmering simulated Seattle, as a Windows 2000 logo whizzed around it to dizzying effect. For those watching, the inference was clear: Microsoft, Seattle and the shape of the future were inseparably interfused.

Microsoft was but the most high-profile of the advanced technology companies that had been a catalyst for Seattle's substantial and sustained economic growth and rapid development over the preceding decade. By the end of the 1990s the Seattle region was ranked fifth in the nation for concentration of advanced technology businesses, generating over $35.9 billion in sales in the year 2000, and had an employment growth rate twice the national average throughout the 1990s.[1] In addition to Microsoft, Seattle found itself home to Abode Systems Inc., Nintendo America, RealNetworks, Amazon.com and around 2,500 other software-oriented firms. For a city previously reliant on

Figure 1: The 'edge city' by night – Microsoft's 'campus' headquarters

the fortunes of the local aeroplane manufacturer Boeing (a state of affairs that saw it particularly hard-hit during the deep recession in the US economy in the early 1970s), this diversification into advanced tech industries such as computer software, microelectronics, robotics and biotechnology ensured that Seattle was one of the main beneficiaries of 'the new economic wave' that had also served to recalibrate the extant hierarchy of US urban centres. As Sharon Zukin notes, the 'regional cleavages' in recent national patterns of economic development had seen many of the older industrial cities of the Northwest (for example, Detroit, Philadelphia and Cleveland) suffer greatly, while cities in the south and west (for instance Denver, Atlanta and San Francisco) experienced rapid expansion and economic growth fuelled by their investment in the advanced services sector.[2] Indeed, as historian Carlos Schwantes points out in his history of the Pacific Northwest, Seattle, at the heart of a region characterised for most of its history as an American Hinterland, had been transformed from a "backdoors of the continent to [a] global gateway".[3]

Yet within eighteen months of the Microsoft launch, the hubristic spirit it so clearly signified would be rendered hollow as a result of the global collapse in dot-com equities, and its deleterious impact on the advanced technology sector.

In the aftermath of the Nasdaq index crash in March 2001, Seattle, a city that had spent much of the 1990s developing a reputation as a hub for advanced technology industries, and as a mecca for those seeking to add their name to the fabled roster of youthful dot-com millionaires, saw much of the lustre disappear as the media began to report unprecedented job losses in the software sector, and a substantial decline in advanced technology employment more generally. By January 2003 the *Seattle Times* had declared "So Long Seattle: More People Are Moving Away", and pointed out that an economic downturn had left the Seattle area with some of the nation's highest jobless rates and the prospect of its first population decline since the 1970s.[4] Clearly, the dramatic impact of the financial slump was seen as presaging the 'end of an era' for the city. In this chapter I take a closer look at the 'prelapsarian' Seattle as advanced technology centre, and consider some of the ways in which it was represented in fiction, on film, and in the American print media in the 1990s. In particular, I seek to problematise the notion that the dramatic dot-com slump of 2001 represented the straightforward shattering of a preceding narrative of giddy economic optimism, and suggest instead that it merely brought into high relief the pre-existing tensions in Seattle's identity as a national symbol of urban, technological progress.

SPACE-NEEDLE VISIONS

For all the advanced technological razzmatazz that surrounded Microsoft's launch of Windows 2000, there was something perplexing about the vision of Seattle's technological future being embodied, as it quite palpably was, in the form of the Space Needle. Constructed in 1962, the Space Needle had been the centrepiece of a World's Fair intended by the U.S government to convey "the country's achievements in science and space ... [and its] continuing affluence and technological advance".[5] A symbol of local and national confidence and optimism, the Space Needle encapsulated the spirit of a period in which America could conceivably envisage the future as comprising 'more of the same'. For instance, writing about the Seattle World's Fair, historian John M. Findlay points out that "the future in 1962 appeared as a richer, easier, and speedier version of the present – not as one in which cultural values and social relationships had changed, and not as one in which there existed significant limitations to America's global power and natural environment".[6] In choosing the Space Needle as the key icon for its vision of technological progress and innovation, Microsoft had selected the symbol of a decidedly anachronistic-looking future.

If Microsoft had been alone in its invocation of the Space Needle as an icon of the future, then the event could be dismissed as an isolated aberration, a high-tech doodle for shareholders and interested parties, with little in the way of wider representational gravitas. Yet this was not the case. During the 1990s Seattle found itself the subject of a number of profiles in the American media which sought to situate the city's identity as a site for advanced technology industries within a narrative of national innovation, progress and economic development. Such accounts saw in Seattle's urban vitality and economic vibrancy a model for a renewed national well-being, in so doing reiterating the fact that America's cities have long functioned as focal points for broader discourses of national economic and social progress and performance. That these accounts sought invariably to reproduce the Space Needle as the totemic visual motif of Seattle's, and thus in some sense the nation's, vitality and health, is worthy of further consideration. I want to take a close look at three such accounts of the city from the period. All are from national magazines and periodicals, and can be seen to comprise some of the most extensive and detailed discussions of Seattle's advanced technology industries carried by the mainstream print media in the 1990s. As such, they can be said to represent significant contributions to a public commentary on the city's identity as an innovative, future-oriented city.

In July 1991, *Lear's* magazine published a seven-page profile of Seattle, written by Mary Bruno and entitled "Seattle Under Siege: The last best city in America fights for its soul".[7] The title set out the article's central concern, which was to position Seattle within a narrative of national urban well-being, and its dramatic portent was indicative of the tenor of the piece as a whole. The 'siege' to which the title referred was, the article stated, the result of "disillusioned urbanites from Burbank to Brooklyn ... descending [on Seattle] in droves". As if that assertion was not sufficiently explicit, the article concluded with the claim that "Seattle has a chance to be a city of the future ... New York is over the edge, L.A. is over the edge. Seattle is one of the last places that has a chance to make things work".[8] *Lear's* stated that what underpinned Seattle's opportunity to 'make things work' was the robust economy, which the article went on to describe as the city's 'trump card':

> The city was too geographically remote and environmentally conscious to ever compete effectively as a manufacturing center. But with today's focus on clean technologies and Pacific Rim trade, Seattle makes perfect sense.[9]

Written in 1991 at a time of nation-wide recession, the attraction of Seattle's robust economy seemed a logical and inevitable one – the bright star in an otherwise gloomy sky.[10] Yet the 'logic' of Seattle lay not simply in its economic well-being, but rather in how that well-being took on symbolic meaning. Thus it was not only that Seattle's preponderance of advanced technology industries was seen to give the city a particularly advantageous position within an emergent paradigm of economic development. Rather, it is how that identity as an advanced technology centre was seen to impact upon the affective and experiential 'economy' of the city. Reflecting the identity of the magazine in which it appeared – this was not a data-driven journal of technology, but a lifestyle periodical – the article's interest in the 'soul' and the 'sense' of the city seems entirely apt. What being a "city of the future" entailed was the hope of Seattle *retaining* its size, scale, the discrete identities of its various neighbourhoods, and its moderate downtown density. The sprawl of Los Angeles and the sheer magnitude of density in New York were placed in opposition to a Seattle which was seen as "the perfect compromise", a phrase which spoke not only to the city's intermediary identity, but also to the notion of inhabitants choosing pragmatically to 'settle' for Seattle.[11]

The key image used to convey a 'sense' of the city was the Space Needle, a photograph of which dominated the first two pages of the *Lear's* article. Accompanied by the caption "Seattle Under Siege" the photograph of the Space Needle took on a synecdochical function – the Needle *was* the city. On the one hand, the Space Needle's identity as the symbol of a time in which the future appeared a "richer, easier, and speedier version of the present" worked to exemplify the article's desire to valorise a continuum of social and cultural ideals – in other words, what was hoped for was that Seattle will 'remain the same', all the while underpinned by a buoyant advanced technology economy. However, there was also a discernible militaristic undertone to the combination

Fgure 2: Space Needle visions: the view of downtown Seattle from Queen Anne Hill

of image and text utilised by the *Lear's* article. The Space Needle's inhabitable top section was made to resemble a 'city in the sky', detached and remote, yet at the same time vulnerable on top of its precarious perch. As Mike Davis argues in *City of Quartz*, the 1980s and 1990s witnessed the emergence of "'fortress cities' brutally divided between 'fortified cells' of affluent society and 'places of terror' where the police battle the criminalised poor".[12] Written at the time of a highly publicised crime-wave in New York, Seattle as Space Needle was clearly conceived and composed to resemble a fortress against the 'siege'. Implicit in this portrait of the city was the notion that what made Seattle important was precisely the fact that such 'threats' to the city were being portrayed as potential, rather than actual, and therefore the city remained distinct from New York and Los Angeles – deemed already 'over the edge'.[13]

A comparable anxiety over Seattle's future could be found in *Newsweek's* May 1996 cover story on the city, which asserted that "Seattle still worries about its future; it is shadowed by its very success, haunted by the fear that stalks every city west of the Mississippi, of turning into another Los Angeles".[14] Published five years later, the article employed none of *Lear's* militaristic rhetoric, but sought similarly to situate Seattle within a nationalised narrative of inter-urban competition and comparison, prefacing its profile of the city with the declaration "Step aside, New York, L.A., Washington".[15] Like *Lear's*, the article described the city attracting incomers from other more established urban centres, stating that it "seems like everyone's moving to the city of Microsoft".[16] Seeking to situate this phenomenon with a social and cultural context, the article stated that:

> The reality of post-industrial society is that power lies with the people who tell the rest of the world what to think. Since the invention of television, that power has been invested in New York, Los Angeles and Washington. But if – as Microsoft chairman Bill Gates seems to believe – it's now up for grabs in the demotic babble of the Internet, then Seattle may be the city to grab it.[17]

Clearly, there were a number of assumptions underpinning this statement, not least the assuredness about the United States' recent history as 'world power', locatable within dominant cultural and political institutions, and capable of being sited quite confidently and tangibly within specifiable large urban centres. Secondly, despite the fact that advanced technologies were identified as posing a challenge to the established order of things – 'it's now up for grabs' – Seattle became the means by which to resituate and re-establish 'world power' once again within an American urban setting. Thus the article sought to generate a sense of reassurance, not only that the United States would continue to play a leading role in the future of advanced technology-oriented economic development, but that this mode of development would continue to organise itself in relation to the nation's large urban centres, and therefore serve to perpetuate what Donald Lyons and Scott Salmon have termed the American "urban hierarchy" – a range of key US urban centres playing pivotal roles within a globalised economic system.[18]

Yet if *Newsweek*, like *Lear's*, proposed Seattle as the city of the future, it is worth asking precisely what the magazine suggested that future would look like. What is clear is that the vision of the city was far from that of the post-industrial city as fragmented, heterogeneous, 'apartheid' urban space as offered by writers on Los Angeles, such as Edward Soja, Mike Davis and Mike Dear. Rather, like *Lear's*, the article prefaced its text with a photograph of the Space Needle, this time against a shimmering night time downtown cityscape, and accompanied by the caption 'The Future Looks Bright'. The article as a whole reflected the photograph's ideological investment in the 'future city' as a continuation of the physically dense, spatially integrated, bounded city of the past. Indeed, it is clear that despite the article's discussion of post-industrial society, it was the urban paradigm provided by the unitary industrial city that was being drawn upon here, and served to generate a strangely nostalgic retro-futurism. If, as Findlay suggests, the Space Needle is the icon of a time when the future did not yet connote "significant limitations to America's global power", then its invocation in *Newsweek* served to generate a seductive vision of Seattle as an alternate vision of America's urban future – to co-opt an evocative phrase from Mike Davis's *Ecology of Fear*, "not so much the future of the city as the ghost of past imaginings".[19]

The tension and anxiety over the fragility and precarious nature of the future that Seattle was seen to represent also infused the cover story on the city in the *Christian Science Monitor*. The newspaper 'discovered' Seattle rather later than *Lear's* and *Newsweek*, producing its cover story on "Seattle's rise as the capital of the New Economy" near the end of the decade in November 1999.[20] Yet despite the period of eight years separating the article from the original piece in *Lear's*, there was a remarkable affinity in tone and theme between the two accounts of the city. For example, the article suggested that:

> In the last third of the twentieth century, Seattle has metamorphosed from an unpretentious city … to a pulsing urban mecca. A month before the new millennium, the influx of new people, new dreams – and mountains of new wealth – has recast Seattle into a prototype of the New Economic Metropolis.[21]

Going on to describe Seattle as "America's new Oz", the article noted that whilst "Boeing is still the single largest employer … software companies and Internet firms make it a New Economy hub".[22] The article repeated *Lear's* and *Newsweek*'s invocation of the prototypical – in other words, that Seattle represented a new urban paradigm, driven by advanced technology growth, and situated quite clearly within a narrative of national progress and economic vitality. The article also incorporated a requisite photograph of the Space Needle, placed in a panoramic shot that also included the city's cluster of downtown skyscrapers. The result was to reinforce the association between Seattle's vital economic growth and the future health of the central city.

However, other sections of the *Christian Science Monitor* article displayed less assuredness concerning the certainty of the city's rosy future. For example, the article stated that the city was characterised by a "civic dichotomy", noting that Seattle is seen to be "making its way in the world with two distinct personalities: one devoted to leading the planet toward the promise of the New World Economic Order (while itself becoming as last cosmopolitan), the other committed to preserving the elite Northwest lifestyle and small-town feel".[23] Indeed, it was precisely the foregrounding of these contradictory tendencies that characterised all three articles' representations of Seattle. Each article professed to catch the city on the cusp of a similarly momentous transformation – for *Lear's* Seattle was a city under siege, for *Newsweek* it was in danger of becoming a victim of its own success, for the *Christian Science Monitor* a city of "civic dichotomy". On the one hand, all three articles manifested a deep imaginative investment in the notion that Seattle could avoid the fate of other American cities, and retain its identity as an economically vibrant, safe, dense urban centre that functioned as a site of work and leisure. On the other hand, the articles were also wedded to the idea of American national economic advancement, as exemplified by the industries of advanced technology with which Seattle was so closely associated. Rather ironically, these advanced technologies were depicted *both* as the means by which Seattle could avoid the painful social and economic consequences of de-industrialisation that had brought economic decline to older urban centres, and therefore served as an emblematic site of national economic well-being *and* at the same time represented the catalyst for overpopulation, urban sprawl, and the host of social ills outlined by the *Monitor* as "more drugs and prostitution, more housing problems [and] new assaults on the environment".[24]

With this is mind, the recourse to the image of the Space Needle could therefore been seen as a means by which these articles attempted to negotiate such contradictions, and one that made perfect sense in relation to their projected readership demographics. All three publications were aimed at affluent, educated and predominantly middle-aged readers, and it is perhaps no coincidence that the median readership age for these publications correlates to individuals who would have been children at the time of the Seattle World's

Fair in 1962.[25] In other words, this is a demographic for whom the future epitomised by the Space Needle – a smoother, faster version of the present in which America's continuing influence and technological advancement seemed assured – would have represented the 'ghost of past imaginings'. As Peter Williams and Neil Smith have pointed out, compared to the early 1960s, subsequent discussions "of the future of the city are more cautious, less speculative, and suffer less from the naïve linearity of extrapolative futurism". They note that "the future is treated as more and more contingent".[26] The retro-futurism depicted in these journalistic accounts of Seattle reproduced just such wariness of vision, whilst also engendering a certain nostalgia for a period in which high-technology represented the unlimited potential of the United States as a global power. As Robert Beauregard points out, the disintegration of that narrative of infinite national progress has often found itself reproduced in the metaphor of urban decline. As he states "the fears and anxieties engendered by a progressively precarious world are thereby focused on the city".[27] With this in mind, it could be argued that the Space Needle visions found in these journalistic accounts presented Seattle as an imaginary space for wresting a seemingly long-lost urban future from the grip of current dilemmas.

SEATTLE'S NEW TECHNOLOGY PIONEERS: MICHAEL CRICHTON'S DISCLOSURE

Representations of Seattle as a centre for advanced technology industry were not restricted to the American print media in the 1990s. Although journalistic accounts suggested a particular agenda in positioning the city within a nationalised narrative of economic progress and advancement, contemporary fiction writers also displayed an interest in exploring the issues raised by Seattle's newly recognised role as a crucible for advanced technology. Among the most high profile of these works of fiction was Michael Crichton's novel *Disclosure* (1994). Over the last thirty years Michael Crichton has emerged as one of America's best-selling popular novelists, beginning with *The Andromeda Strain* in 1969, and including other notable successes such as *Congo* (1980), *Sphere* (1987), *Jurassic Park* (1990), *Rising Sun* (1992), *The Lost World* (1995), all of which have also been made into highly successful motion pictures.[28] Crichton has also directed a number of movies, including *Westworld* (1973) and *Coma*

(1978), as well as being the creator and executive producer of the Emmy Award winning prime-time medical drama television series *E.R.* (1994–). Crichton's influence on American popular culture, reflected only in part by his oft-cited status as the inventor of the modern "techno-thriller", makes his creative output worthy of closer consideration.[29] Moreover, Crichton's work, characterised by a concern with the ways in which advanced technological and scientific development intersect with contemporary social issues, has been frequently accompanied by commentaries which proclaim the timely nature of the novel or film in question.[30] For example, his Seattle-set thriller *Disclosure* was one of the best-selling novels of the early 1990s, as well as a book that generated extensive commentary in the popular press at the time of its publication. This was due in large part to the book's reversal of the standard depiction of sexual harassment in the modern office, and which made the sexual predator in question an attractive young woman and the victim a middle-aged family man. This contemporary perspective on office sexual politics was combined with a similarly 'zeitgeist' location – namely a Seattle-based computer software development company. Ostensibly the backdrop for the tale of advanced technology machinations, Seattle was nevertheless mobilised by the novel in a manner which connected it in interesting ways with broader questions of urban representation, urban renewal and the city's role as a symbol of national economic well-being in the 1990s.

Disclosure tells the story of Tom Sanders, a senior executive for the Seattle software firm DigiCom, who is passed over for promotion by Meredith Johnson, a new female executive who also happens to be Tom's former lover. After Tom rejects Meredith's sexual advances during an after-work drink, he finds himself accused of sexual harassment. Meredith demands Tom's transfer from DigiCom's Seattle headquarters, thereby denying him the millions of dollars he would have made when DigiCom floated on the stock-market, as well as putting a severe strain on relations with his wife and children. Tom refuses to leave, and instead decides to fight the charge, thus embroiling himself and his family in a potentially high-profile lawsuit. With the help of an anonymous company insider Tom finds out that the sexual harassment charge is simply an elaborate smoke screen for Meredith's primary objective, which is to cover-up her serious mishandling of activities at DigiCom's production plant in Malaysia. The novel ends with Meredith's public humiliation as the truth is revealed at a company meeting, and with Tom's reinstatement as division manager at DigiCom.

Disclosure depicts a Seattle populated by middle-class, white-collar workers, reflecting the book's focus on machinations within the advanced service sector. Thus the beginning of the novel describes Tom's journey to work from his home on Bainbridge Island, an affluent suburb located a commuting distance away from the city of Seattle in the Puget Sound. Tom is subsequently described rushing to make the commuter ferry that will take him on the thirty-five minute scenic journey to DigiCom's offices, in the heart of downtown Seattle. It is important to note that Crichton does not simply generate an imaginary location somewhere in the city, but instead selects an actual location, namely Pioneer Square.[31] Crichton writes:

> DigiCom was located in three different buildings around historic Pioneer Square, in downtown Seattle … Around Pioneer Square were low-rise red-brick buildings built in the early years of the century, with sculpted facades and chiselled dates; these buildings now housed trendy architects, graphic design firms, and a cluster of high-tech companies.[32]

Crichton's depiction of Pioneer Square reflects the author's fabled reputation for meticulous research – the Square was indeed a key site for the relocation of some of Seattle's more select white-collar businesses during the 1990s. But the choice of Pioneer Square also mobilises a number of symbolic associations that are particularly pertinent. Pioneer Square, as the name implies, harks back to Seattle's original settlement in the 1850s, a time when the Puget Sound region was at one edge of "the double frontier" – in the sense that it gained white inhabitants before the Great Plains further to the east. In setting out beyond the edges of existing settlement two hundred miles south-east in Portland, the founders of Seattle were thus deemed "pioneers".[33] Yet the function of 'pioneer' in relation to the theme of the novel suggests two further connotations. As David Wrobel, a historian of the American West notes, during the late-twentieth century the term frontier became detached from its geo-historic referent to become a generalised "metaphor for promise, progress, and ingenuity".[34] Indeed, during the 1990s metaphors of 'frontier' and 'pioneer' found extensive application in relation to the industries of advanced technology, thus making Crichton's use of Pioneer Square particularly apt. Moreover, such metaphors continued to function in association with a narrative of national promise, progress and ingenuity. For instance, Robert X. Cringely's seminal history of the personal computer industry, entitled *Accidental Empires* (1996), provides a striking example of the ways in which accounts of advanced technology industries lent themselves to narratives of the frontier, and consequently into contemporary formulations of national progress and development. In an early passage in the book Cringely describes the way that such narratives structure themselves:

> America was built on the concept of the frontier. We carved a nation out of the wilderness, using as tools enthusiasm, adolescent energy, and an unwillingness to recognise limitations. But we are running out of recognised frontiers. We are getting older and stodgier and losing our historic advantage in the process. In contrast, the PC business is its own frontier, created inside the box by inward-looking nerds who could find no other acceptable challenge in the adult world. Like any other true pioneers, they don't care about what is possible or not possible; they are dissatisfied with the present and excited about the future.[35]

Written in the mid-1990s, Cringely's account suggested that the PC business could provide a 'new frontier' for a nation that had exhausted other options, or perhaps more pertinently, for an exhausted nation that had few other options. In this way, the PC business represented an opportunity for Americans to be pioneers again, to be the first, and to be number one. Indeed, he noted that the industry included its own built-in 'natives' to conquer, in the shape of industry competitors – other nations, who could not grasp the frontier with quite the success of the Americans. Examining 'the Europeans' Cringely suggested that their success on the multi-media frontier was hampered by being "obsessed with style" and producing "high-tech toys that look pretty, cost a lot, and have such low performance".[36] In a similar vein, Cringely suggested that "[the Japanese] can't take over" the industry either because "they are too businesslike, too deliberate, too slow".[37] Thus Cringely's account was firmly in keeping with the way in which depictions of Seattle in the aforementioned periodicals worked to situate the city as an advanced technology nexus within a narrative of economic well-being – as, in the words of the *Christian Science Monitor*, 'leading the planet toward the promise of the New World Economic Order'.

This mobilisation of the frontier as a symbolic resource within American culture has been noted by a number of scholars. In *Gunfighter Nation* Richard Slotkin traces the evolution of the frontier myth in American culture from 1890 onwards, in other words, from precisely the moment when "the landed frontier of the United States was officially declared 'closed'".[38] From that moment on, Slotkin argues, the frontier became "primarily a term of ideological rather than geographical reference".[39] Slotkin examines the way in which the frontier myth worked its way, in mutated form, into a whole range of ideologically charged scenarios from Korea and Vietnam, to the Reagan administration's policies towards Central America and George Bush's "War on Drugs".[40] In such scenarios, the frontier myth functioned as a rallying cry to gather forces against a 'savage' enemy, whether it was the Viet Cong or General Noriega. In this case, the 1990s multi-media frontier became the site of America's potential resurgence within the context of profound socio-economic restructuring, and the global competition for markets, jobs and revenue.

Yet if Crichton's choice of Pioneer Square functioned as an adroit way of drawing upon Seattle's wider representation as an advanced technology centre in the 1990s, and thus as a key site for national narratives of "promise, progress and ingenuity", there is another sense in which the location had an important symbolic resonance. Pioneer Square, described by the novel as "low-rise red-bricks" housing "trendy architects, graphic design firms, and a cluster of high-tech companies" was one of the key sites of downtown urban renewal in Seattle in the 1980s and 1990s. As Neil Smith points out in *The New Urban Frontier*, popular narratives of urban renewal, circulated by the national media in the 1980s and 1990s, and exemplified in lifestyle magazine profiles of warehouses and wharves converted for 'stylish middle-class city living', and the proliferation of gentrified markets and waterfront districts, found a number of applications for the frontier motif. Smith states that "in the language of gentrification, the appeal to frontier imagery has been exact: urban pioneers, urban homesteaders, and urban cowboys became the new folk heroes of the urban frontier".[41] As he notes "the term 'urban pioneer' is therefore as arrogant as the original notion of 'pioneers' in that it suggests a city not yet socially inhabited".[42] Moreover, it also draws upon a similar sense of "the forging of the national spirit".[43] Smith writes:

An equally spiritual hope is expressed in the boosterism which presents gentrification as the leading edge of an urban renaissance … the new urban pioneers were expected to do for the flagging national spirit what the old ones did: to lead the nation into a new world where the problems of the old one were left behind.[44]

In situating the story of advanced technological machinations in Pioneer Square, *Disclosure* therefore drew heavily upon Seattle's symbolic location on a contemporary "double frontier" in the 1990s. On the one hand, the Square reflected the city's status as a key site for advanced technology development, on the other it represented the type of downtown urban development closely associated with the complex processes of socio-economic restructuring within advanced capitalist economies.[45]

It is worth pointing out that for all the 'zeitgeist' sexual politics and advanced technology gadgetry, *Disclosure* offered, at heart, a timely tale of middle-class status anxiety. The 'fear' that drives the novel is that of Tom losing his job, and thus his privileged middle-class lifestyle, as well as the break-up of his nuclear family. Indeed, it is interesting to note the difference in the way that this central structuring theme is handled in the novel and in the 1994 film adaptation (itself a highly successful work of fiction, taking over $83 million at the US box office).[46] Whilst the theme is somewhat submerged within the complex array of plot lines in Crichton's typically dense and detailed prose, the film's attenuated story foregrounds it much more. For example, the aforementioned opening scene of the novel, depicting Tom's journey to work in Pioneer Square, is repeated in the film, but with a few significant differences. In the film Tom (played by Michael Douglas) is driven to the ferry by his wife Susan, rather than making his way there alone. Firstly, we are given a long shot of the Space Needle through a copse of trees in front of their house. Not mentioned in the novel, the film's depiction of the Space Needle not only locates us as viewers in relation to the city of Seattle, but also fits nicely with the film's more cautious narrative of technological futurism. Indeed, the Space Needle sets the tone for the ensuing conversation between Tom and Susan, again not featured in the novel. Tom informs Susan of the developments within DigiCom that are likely to lead to his lucrative windfall from the proposed merger. Susan's response is to recount an old adage gleaned from her mother, cautioning "don't climb up there next to God – he might shake the tree". We do not get to witness Tom's response, as the scene quickly cuts to his departure on the commuter ferry, but the weight of portent is clear. Moreover, the next scene on the ferry serves only to reinforce the sentiment. We are shown Tom chatting to a man who is obviously a fellow commuter, familiar to him from repeated ferry journeys. The man is identified as a recently laid-off employee of IBM, one of the giants of the American computer industry, but also a company that the knowledgeable viewer would recognise as one of the 'losers' in the cut-throat wars to establish dominance in the US software industry in the late 1980s.[47] The nameless ex-IBMer embarks upon a monologue as Tom talks on his mobile phone:

> Twenty-eight years with IBM … did I ever tell you what they told me? I was surplused. You ever hear that word? If they wanted a euphemism they should have said sodomised. You just don't see it coming, you're just going right along, and then one day there's no room … boom! "no more room for you!" Smaller, faster, cheaper, better…[48]

Tom tries to curtail the man's ranting by passing him the phone number of Cindy, his secretary, in order to see whether a position may be found for him at DigiCom. Instead, it only sparks another monologue on the changed nature of the working environment, the ex-IBMer stating that "Cindy … pretty name. Used to be you could have fun with the girls, nowadays she probably wants your job." The ex-IBMer not only proclaims his own redundancy but also compounds his status as a superseded model by displaying archaic work-place behaviour and attitudes.

The figure of the ex-IBMer is not in Crichton's novel but the role he plays in the attenuated story of the movie is clear. Firstly, by not naming the ex-IBMer the film gives him the status of an emblematic figure – less a fully-rounded character than an 'everyman' – a warning to Tom about the tenuous status of his own lofty perch. Moreover, coming immediately after Susan's warning about climbing "up there next to God", the ex-IBMer's monologue compounds the sense of anxiety generated by the film. In this way the film, much more explicitly that the novel, engages with a broader discourse about the fragile nature of employment within the advanced service industries.

Figure 3: High Anxiety: Tom Sanders (Michael Douglas) gets to grips with Meredith Johnson (Demi Moore) in *Disclosure*.

As Geoff Mulgan points out, despite the fact that 1980s and 1990s popular culture "invariably situate[ed] information technology in downbeat, insecure futures dominated by monstrous global corporations: the doomed androids of *Blade Runner*, the decaying anarchy of cyberspace in cyberpunk literature" what was really at stake here, and what generates the thematic core of the film of *Disclosure*, is a much more prosaic "shifting balance of power".[49] As Mulgan states:

> Today's main victims of technology are not only the thousands of bank tellers and shop assistants but also layers of middle management whose role as conduits of information up and down organisational hierarchies has become redundant.[50]

Indeed, Mulgan's argument finds support in the work of Barbara Ehrenreich, a writer who has become an influential chronicler of white- and blue-collar employment patterns in the US. Ehrenreich's conceptualisation of a new middle class or what she calls a "new professional-managerial class", is one very much associated with the rapid growth of white-collar employment in service industries and advanced technology and financial sectors.[51] Moreover, the volatility of such employment, dependent as it is upon the fluctuating fortunes of global markets for labour and services has, argues Ehrenreich, engendered a workforce characterised by a profound 'fear of falling'. Outlined in her study of the same name, Ehrenreich's notion of the 'fear of falling' is used to characterise what she sees as the increased anxiety of the middle class as regards misfortunes that "might lead to a downward slide".[52] Ehrenreich's book investigates the middle class's fear of losing status, and social and economic control, in an era of greatly increased fiscal and societal uncertainty. Her notion of the 'fear of falling' is based upon the premise that the professional middle class's status *as* middle class is never secure: its "only 'capital' is knowledge and skill, [which] unlike real capital … cannot be hoarded against hard times, preserved beyond the lifetime of the individual".[53] As she states, "it must be renewed in each individual through fresh effort and commitment", and no one "escapes the requirements of self-discipline and self-directed labour", whilst aware that they are always at the mercy of volatile economic fluctuation.[54] Ehrenreich's argument also finds affinity in the work of Sharon Zukin, who suggests in *Landscapes of Power* that one of the great transformations in American society in recent years has been the disappearance of "the industrial prosperity that was guaranteed by US dominance in the world economy and a shared middle-class way of life". She notes that in their place "rise alternate images of great wealth, insecurity, and fragmentation".[55]

It is an obvious point to make that as a 'techno-thriller' it is unlikely that *Disclosure* would portray the impact of advanced technology upon Seattle with the nightmarish dystopianism of, say, science-fiction. However, it is possible to argue that if the socio-spatial restructuring of Los Angeles has lent itself readily to extrapolation and exploitation in films such as *Blade Runner* and *Strange Days*, then *Disclosure* suggests a quite different order of relationship between the city as material environment and as imaginary space. If depicting the impact of advanced technologies on Los Angeles has been accompanied more often than not by what Liam Kennedy calls "the theme of impending apocalypse", then *Disclosure* presents a quite contrasting role for Seattle.[56] In particular, the film version of *Disclosure* narrativised the sense of anxiety apparent in a number of journalistic accounts of Seattle's advanced technology industries in the 1990s. It drew upon Seattle's portrayal as a city 'on the edge' – not only in terms of its significant concentration of advanced technologies, but also the fragile nature of its status as a city that had a chance to avoid the fates of other large urban centres. In other words, *Disclosure* mobilised this broader discourse on Seattle in order to generate the perfect location for a parable of middle-class anxiety in a progressively precarious world.

DOUGLAS COUPLAND'S MICROSERFS AND THE 'TECHNOLOGY CORRIDOR'

An intriguing way in which it is possible to illustrate the particular symbolic investments apparent in all the preceding accounts of Seattle's identity as a key centre for advanced technology industries is simply to draw attention to the disparity between the socio-spatial organisation of the Seattle they delineate, and the actuality of the city's technological development in the 1990s. In other words, reference to the actual impact of advanced technology industries on the city provides a means of highlighting the complex relationship between the cultural work of representation and the actual city. For example, consistent across all the representations discussed thus far has been the promulgation of a traditional suburb/city-centre dynamic. Tom Sanders' commute in *Disclosure* replicates the familiar journey from home on the periphery of the city to work in the metropolitan core, thus reflecting the established post-war trend towards residential suburbanisation, and the centralisation of corporate headquarters. Similarly, the journalistic accounts of Seattle suggest an investment in the city's retention of this "dialectic of spatial centralisation and decentralisation", with a vibrant and concentrated urban city core.[57] Indeed, in these accounts, advanced technology industries act to underpin Seattle's identity as a key site of urban renaissance, with gentrified spaces housing "trendy architects, graphic design firms, and a cluster of high-tech companies".

However, the overwhelmingly *symbolic* appeal of situating Seattle's technological future in the downtown, as outlined earlier, becomes all the more obvious when one considers where Seattle's technological future was actually being forged in the 1990s. By far the most important and substantial area for the location of advanced technology industries in the Seattle area throughout the 1990s was the appropriately named Technology Corridor, a ten-mile path north-west of the city in South Snohomish County, stretching along Interstate

405 between the towns of Bothell and Everett. The Corridor attracted more than "250 electronic, telecommunications and computer businesses [that] cluster on quadrants, campus-like neighborhoods, [with an estimated] 50,000 jobs".[58] Amongst the employers located in the Technology Corridor were Honeywell, Advanced Technology Laboratories, Boeing and Microsoft.[59] In addition, the I-90 Corridor, Bellevue (which, reflecting its past as a suburb, was once called 'Boeing's bedroom'[60]) and South Center-Kent Valley also represented the ex-urban "heart of the region's computer software industry".[61]

The journalist Joel Garreau famously coined the phrase 'edge city' in the early 1990s as a means of describing such patterns of ex-urban development.[62] In fact Garreau actually identified the Seattle area's Bellevue-Redmond, the Technology Corridor, the I-90 Corridor, and South Center-Kent Valley as "edge cities" in his influential study of the phenomenon. Garreau used the phrase 'edge city' to describe what he termed as a "third wave" of urban development in the United States, a period in which Americans "have moved our means of creating wealth, the essence of urbanism – our jobs – out to where most of us have lived and shopped for two generations".[63] These new locales are deemed "edge" by Garreau, in part because of their geographic location – they "ris[e] far from the old downtowns, where little save villages or farmland lay only thirty years before", but also because of their status as a "psychological location" on the cutting edge. Garreau asserted that these developments were the future: "how cities are being created", reflecting the "profound changes in the ways we live, work, and play".[64]

Such 'edge city' development in the Seattle area did not go unnoticed by novelists. In particular, the writer Douglas Coupland provided an extensive exploration of the 'edge-city' of Bellevue-Redmond in his novel *Microserfs* (1995), published only one year after *Disclosure*, but examining a quite different socio-spatial mapping of Seattle as advanced technology nexus. Indeed, Coupland provides an interesting counterpoint to *Disclosure*'s Michael Crichton. In terms of sales and sustained success, Coupland cannot compete with Crichton, yet he represents an influential commentator on contemporary American culture in his own right. Starting with his acclaimed and commercially successful first novel *Generation X* (1991), which portrayed young Americans struggling to cope with a bleak future of diminished expectations of material wealth, Coupland has attracted much critical attention as a self-styled (albeit outwardly reluctant) spokesperson for his generation.[65] Coupland's novels and essays have all drawn upon aspects of popular culture in interesting ways, in particular in seeking to make connections between pop cultural detritus – adverts for GAP clothing, IKEA catalogues, Grateful Dead concerts – and the ability of individuals to make connections and build relationships in a rapidly changing and unstable world. Like Crichton, Coupland's novels have been noted for their emphasis upon 'zeitgeist' thematics, inevitably picked up in critical reviews. Yet whereas Crichton arguably keys into the obsessions of the baby boom generation – as exemplified by the middle-aged, middle-class protagonist of *Disclosure* – Coupland's focus has most usually been upon the next generation – America's twentysomethings, frequently depicted as still struggling to establish themselves in a turbulent world.[66]

The importance of Coupland's *Microserfs* lies largely in its status as the first novel to depict life in a US edge city, and also in the fact that it was the first novel to generate a fictionalised account of the culture of the Seattle software giant Microsoft, whose impact upon the computer industry, and American culture more generally, has been profound. Coupland's *Microserfs* approaches Microsoft from the perspective of a group of young workers at the corporation's headquarters in the 'edge city' of Bellevue-Redmond. The novel is written as the journal of Daniel, a 26-year-old 'bug tester' at Microsoft, and described the changes in the lives of his friends and himself, over a couple of years, starting in the fall of 1993. The 'Microserfs', as Coupland calls them, are shown at the start of the novel working mind-numbing 18-hour days in the Microsoft campus, filling the rest of their waking hours with junk food, TV adverts and e-mail. As Daniel states at one point, "I am 26 and my universe consists of home, Microsoft and Costco."[67] Written with scathing wit, Coupland charts the Microserfs' growing and weary rejection of Microsoft and the societal aspects of its corporate culture, and describes their decision to leave the safety of the campus in order to start up their own software company, named *Interiority* (signifying their desire to find more ephemeral and spiritual aspects of life). [68] Daniel and his friends not only flee from Microsoft, but from the edge city of Bellevue-Redmond altogether, as they go on to create and market their software company in California's Silicon Valley.[69]

Although only the first hundred pages of *Microserfs* portrays life in Bellevue-Redmond, in that space Coupland managed to generate a detailed and vivid description of life in the 'edge city', and one that provided a distinct contrast to the depictions of Seattle as advanced technology centre offered by national periodicals and Crichton's *Disclosure*. For example, in an early passage in the novel, Daniel describes the group house in which he lives with other Microsoft employees, a short drive from the Microsoft 'campus' (a real term used to describe the headquarters, and repeated by Coupland in the novel) in Redmond:

> Living in a group house is a bit like admitting you're deficient in the having-a-life department, but at work you spend your entire life crunching code and testing for bugs, and what else are you supposed to do? Work, sleep, work, sleep, work, sleep. I know a few Microsoft employees who try to fake having a life – many a Redmond garage contains a never-used kayak collecting dust.[70]

The group house, in keeping with the 'campus' analogy of the Microsoft headquarters, is an extension of the college students' house. Indeed, Daniel describes the "McDonald's-like turnover in the house", a reference to the temporary, transitory nature of an abode with little of the investment, fiscally and emotionally, of the Bainbridge Island home of *Disclosure*'s Tom Sanders.[71]

The entire section of the novel that deals with life in the Bellevue-Redmond edge city is teeming with disposable detritus and litter: Skittles candy packets, computer magazines, Costco boxes and post-it notes. In one notable passage Coupland has one of the characters reflect that "the 90s will be the decade with no architectural legacy or style … code is the architecture of the 90s".[72] Correspondingly, Coupland's prose is careful to construct little of substance in terms of 'edge city' architecture – everything is part of a shrink-wrapped, interchangeable, IKEA landscape. In his words, everyone is "shopping for the same 3bdrm/2bth dove-gray condo".[73] Indeed, Coupland succeeded in generating very little sense of the 'edge city' as a tethered location, either geographically or psychically – instead he mapped a world of objects and relationships characterised by the theme of in-built obsolescence. In fact, one of the reasons why *Microserfs* moves the characters and the story away from the Seattle edge-city and towards California after a hundred pages of the novel is precisely because Coupland has worked so effectively to produce a highly pressurised, claustrophobic, monotonous, oppressive, regimented landscape, which is defined only by its vectors from home to the Microsoft campus – to leave Microsoft means inevitably to leave the edge city, since the two are shown to exist in symbiosis.

My point here is not to set up *Microserfs* as offering a more realistic portrayal of Seattle as advanced technology nexus – it is, after all, a fictionalised account of life in the 'edge city,' and ultimately used the terrain as a device with which to explore aspects of the contemporary human condition. However, *Microserfs* does offer an interesting point of contrast to *Disclosure* and the national periodicals' accounts of Seattle's advanced technology industries, not only in its vision of the socio-spatial restructuring apparent in the influence of edge city development, but also in its willingness to engage in the type of conceptual restructuring necessary to address the representational challenges posed by profound socio-economic transformations. What is evident is that *Microserfs* posed a distinct challenge to the dominant orientation of a mainstream media discourse on Seattle's identity as a centre for advanced technology in the 1990s. It flagged up the lacunae in the repeated emphasis upon Seattle's downtown city core as the key locus for the area's advanced technological development. What is clear is that the preceding accounts reflected a distinct ideological investment in the imaginary space of Seattle as the organising site and symbolic repository for national narratives of economic well-being, brought into heightened relief when placed together in such a way.

While the selected texts examined in this chapter provide but one pathway through the range of responses to Seattle's development in the 1990s, they nevertheless represent significant contributions to an evolving discourse which questioned the impact of advanced technology on America's cities during the period. That discourse generated a particular role for Seattle as an emblematic site, and one that drew upon changes and transformations in the city in complex and often contradictory ways. The cautious retro-futurism symbolised in the slender figure of the Space Needle was in line with what Kevin Robins describes as "a defensive response that look[s] to old certainties, securities and familiarities", when faced with the profound socio-economic restructuring of America's urban centres over the last twenty years. By contrast, Coupland's assertion in *Microserfs* that "code is the architecture of the 1990s", suggested a quite different attempt to map Seattle's role within an advanced technology landscape, and to represent and visualise the "edge cities" as emerging urban agglomerations.[74]

CHAPTER 2

SELLING THE 'NATURAL NORTHWEST':
SEATTLE AND THE GREAT OUTDOORS

As mobility and hybridity become watchwords of the way the world now works, tradition-bound, defensive articulations of the region may start to look untenable, but the continuing appeal of regionalism … suggests otherwise, even if – as is often the case – regional traditions are exposed as invented.[1]

Discussing his adopted home city for a 1996 BBC documentary, the writer Jonathan Raban declared that in recent years Seattle had become the "first big city … in the history of big cities … where people have come to the city in order to get closer to nature".[2] Raban's comment spoke to the fact that the 1990s saw Seattle gain particular attention as a result of the fact that it seemed to offer a point of reconciliation in an age-old conflict – namely that perceived to exist between the country and the city. Indeed, as the literary critic Raymond Williams points out, "the contrast of country and city is one of the major forms in which we become conscious of a central part of our experience and the crises of our society".[3] Within the United States, this contrast found its first major formulation in the eighteenth century, as the growth of large commercial cities such as New York, Philadelphia and Boston seemed to increase and amplify the clash of values between urban centres and rural society.[4] The development of the nineteenth-century industrial city served only to widen the divide, as the urban became closely associated with immorality, crime, vice and the corruption of incoming rural youth, whilst the rural remained "the bastion of morality" and religious values.[5] Even as its cities were recognised as sites of learning and communication, American society retained a strong strain of anti-urbanism, grounded in the inherent 'unhealthiness' of the nation's urban centres.[6] Massive urban population growth in the late nineteenth century combined with industrial capitalism to create environmental pollution and city slums. Overcrowded tenements, unsanitary living conditions and factories spewing dirt and grime represented key instances of urban ills. Thus the rise of city beautification societies in a number of the nation's big cities in the early twentieth century, implementing tree-planting initiatives, as well as the establishment of city parks, pointed to the influence of a pastoral view of nature as palliative and refuge from urban ailments.

Yet despite the widespread implementation of such initiatives, urban environmental degradation continued throughout the pre-war period. In *Ecology of Fear* Mike Davis charts the failed attempts of Los Angeles to preserve its natural urban spaces in a chapter entitled 'How Eden Lost Its Garden'. He quotes the warning in 1909 by Charles Mulford Robinson, one of the city's chief 'beautifiers' that "if Los Angeles wavered in its commitment to public space, other 'more beautiful' cities would usurp its destiny". Davis adds, by way of commentary, "was he already thinking of Seattle?"[8] He notes the catalogue of environmental catastrophes over the next thirty years that made obvious Los Angeles's failure to ensure that preservation kept pace with the demands of a rapid population explosion, citing "unspeakable beach pollution … floods and sewage spills", and, in 1943, the city's first smog attack.[9]

Throughout more recent times, Los Angeles' frequent and infamous acrid smog attacks became a key symbol for the chronic environmental problems affecting the sprawling metropolis. High levels of air pollution spoke to the manifest unhealthiness of Los Angeles as an urban area. As Davis notes, "only Mexico City has more completely toxified its natural setting".[10] Moreover, as he suggests, such ecological self-destruction found itself represented in "fictional and nonfictional accounts of imminent ecological collapse".[11] Tracing the theme of environmental crisis through urban disaster fiction, Davis points to the allegorical tenor of tales of ecological self-destruction, as mutant nature took 'revenge' upon callous and profligate humanity. It is worth noting that such tales have by no means been restricted to Los Angeles, even if the city represents their most fervent locale. For example, in *Alligator* (1980) the flushing of a baby alligator down the toilet and into the Chicago sewer system is the start for a monster movie with an oft-cited ecological subtext, as the alligator grows to gargantuan size from eating the corpses of animals who have been the object of suspect hormone experiments. From the perspective of near-future fiction, New York has also found itself the location for powerful images of post-ecological disaster from *Soylent Green* (1973) to the submerged Manhattan of Steven Spielberg's *A.I.: Artificial Intelligence* (2001). Consistent across much of this work is the notion that America's large urban areas exist in an antithetical relationship to the natural world, behaving as voracious and unremitting consumers, polluters and contaminators, and ultimately facing the prospect of reaping all that they have sowed.[12]

Such works of fiction can be understood in large part as fantastical extrapolations from contemporary environmental concerns about the "sustainability of terrestrial ecosystems; total air pollution loadings; contaminants in food and water; and health of the oceans" all exacerbated by the concentrated demands of massive urban populations.[13] Throughout the 1990s the necessity of making America's cities somehow more 'natural' became a matter of increasing concern in academic and popular literature, reflecting the central role played by pollution and environmental degradation in accounts of urban crisis and decay.[14] During this period the city of Seattle found itself mobilised within a range of accounts as the centre of a 'natural Northwest', suggesting its identity to offer an alternate vision of the American city's relationship to its natural surroundings. If, as Alexander Wilson points out, the period saw "images and discussions of nature ... increasingly phrased in terms of crisis and catastrophe" then Seattle's promulgation as a city 'closer to nature' suggested this was by no means the only mode of response to attempts to reconcile the city and nature.[15] In the following sections I look closely at some of the ways in which Seattle came to be represented as a 'natural city' in the 1990s, and, crucially, how that image of 'naturalness' became a key selling point for the city, reproduced across a range of commodity forms. I begin by setting the context for Seattle's 'natural' charm.

SEATTLE'S NATURAL ORIGINS

On one level, the propensity of mainstream media accounts to reference Seattle's natural surroundings is a simple recognition of the city's proximity to a stunning natural landscape.[16] For example, it is easy to find magazine profiles of the city from the 1990s that paid straightforward homage to the city's

Figure 4: Majestic View: Seattle with the snowy peak of Mount Rainier in the background

"sensational setting", "sweeping views" or admired the majestic "fog-bound peaks of the Olympic Peninsula and the Cascades".[17] Yet this commonplace sense of Seattle as an entry point to, or on the doorstep of, 'the great outdoors' is actually underpinned by a notion of nature as 'urban appendage' that speaks to the profound spatial re-orientation or re-organisation of the perception of nature, and which directly coincided with the processes of urbanisation and industrialisation. In other words, it is important to recognise the historical specificity of particular conceptualisations of nature, and to acknowledge the constructed and contingent nature of their formation. Carlos Schwantes points out that Seattle is at the heart of a region that is unthinkable "without its mountains, its rugged coastline, its Puget Sound fogs, its vast interior of sagebrush, rimrock and big sky".[18] Schwantes suggests that the identity of the Pacific Northwest, like few other parts of the United States, "is almost wholly linked to natural setting".[19] Yet for most of the region's history nature has had to fulfil an increasingly uneasy double function. On the one hand, nature helped sell the Pacific Northwest to settlers and investors as an area of abundant, exploitable resources, providing "profits and jobs" in logging, fishing, mining and other ancillary industries. On the other, it has been "revered as a source of aesthetic pleasure and outdoor recreation".[20]

Nowhere has the shifting and divided identity of nature been experienced more acutely than in the region's cities. Initial inhabitants of the cities of the Pacific Northwest sought actually to countermand the region's status as a remote "colonial hinterland" by underplaying their "natural setting".[21] Seattle was originally to be named New York Alki (alki from the Chinook Native American vernacular for 'eventually' or 'by-and-by') whilst "a coin flip in 1845 between a Maine and a Massachusetts native resulted in a city named Portland rather than Boston".[22] The founding fathers of Tacoma "instructed engineers 'to loose sight of the fact that [the town] was as yet a wilderness; to forget the forest that bearded the hillsides; to forget that they were on the frontier, and to anticipate the coming of a city of hundreds and thousands'", an instruction that suggested a decided reluctance to come to terms with the specificity of the region's topography.[23] Indeed, it was only in the 1880s, when the railroad companies sought to create a distinct regional identity in order to attract settlers and investors, that a "'natural' Northwest, a place not only of resources to be extracted, but also of scenery to be appreciated and visited began to be promoted and disseminated".[24] As John M. Findlay points out, "the main purpose for advertising nature [was] to attract people who would help develop the country by exploiting its resources".[25] It was only at this time, and under these conditions, that the natural surroundings were invoked, positively, as a way of *differentiating* the cities of the Pacific Northwest – in such advertising, "even cities became part of nature's Northwest".[26] The speed of this volte-face was such that by around 1890, "so prominent had the natural become in regional thinking that it was casually blended with the urban into a mixture taken not just to rival but to surpass the eastern city".[27]

In stark contrast to the proclamations of nature's abundance so prevalent at the turn of the century, protective environmental legislation from the early 1970s onwards has underlined a growing consensus that the region's nature is "at risk" and sought to emphasise the scarcity of natural resources.[28] Indeed, Schwantes has gone so far as to state that "coming to terms with nature's limitations ... [has] produced ... something of an identity crisis" in the region.[29] This was perhaps been felt most combatively in Seattle, whose status as the "uneasy capital of the timber empire", where "aside from protest demonstrations, one will see neither loggers nor log trucks", pitched it against the (increasingly beleaguered) logging community.[30] Schisms had been caused in part by the fact that, for example, 1997 saw only one active mill anywhere in the immediate Seattle area yet timber remained the third most important sector of Washington State's economy.[31] Local journalist Eric Lucas has cited surveys that demonstrate the consistent "deep and abiding distrust among city residents for logging practices" in the face of industry insistence that "there is not a person in Seattle whose life is not dependent on a healthy timber industry".[32]

The 1990s thus saw Seattle become the site for an increasingly recalcitrant debate about the identity of nature, a debate that was at the same time a manifestation of the changing identity of the city, and more broadly, that of the region. To quote Schwantes, the conflict between "metropolitan dwellers who regard the non-agricultural hinterland as a recreational escape" and the rural population who "wish to use it for its natural resources" became more heated as the region transformed into one that had "more teachers than loggers, and more white-collar professionals than commercial fishermen".[33] It also became more hotly contested as nature tourism continued to exert an influence upon the region's economy, and the realisation grew that "some river canyons may be more valuable for their wild waters than as hydroelectric reservoirs, [and] timbers more valuable as shade for hikers and campers than as two-by-fours".[34]

At a local level, it was not hard to find journalistic accounts of Seattle in the 1990s that made apparent this notion of a changing identity. For example, *Seattle Magazine*, a glossy monthly magazine aimed at the "young, affluent Seattle metro resident", dedicated its July 1999 front cover to profiling what it termed "21 terrific outdoor adventures".[35] In the accompanying feature article, writer Laura Slavik stated that:

> We are at the center of things. It's one of the payoffs of living here. Drive (or paddle) an hour in almost any direction and you arrive somewhere uncivilised and seemingly undiscovered … Each spot, close enough to home but far enough away to feel remote, is guaranteed to keep thoughts of PalmPilots and PCs at bay.[36]

Slavik's portrait of Seattle acted to conjoin the city's identity as an advanced technology centre with an image of the location as an entry point to "the great outdoors". So chic and breezy was the portrait that it seemed almost churlish to point out that the imagined community of PalmPilot and PC users the article addressed was a rather circumscribed portion of the city's population. Yet the fact of the matter is that *Seattle Magazine*'s portrait of the city's relationship to its natural surroundings had a significance that existed way beyond its local readership, and made manifest in the magazine's insistence that being surrounded by nature accorded Seattle the status of being "at the center of things".[37] As Carlos Schwantes enquires in his chapter "Redefining the Pacific Northwest", "how and when did the region's historical identification with nature cease being a liability – the mark of a cultural backwater – and become an asset?"[38]

One response to Schwantes's question relates to the increasing attraction of Seattle's 'natural' identity within the context of an urbanised United States characterised by profoundly unhealthy relationships to their natural environments. If we recognise that concerns about the unhealthiness of urban centres were both a legitimate response to an increasingly widespread awareness of environmental degradation, and also yet another way in which the fears and anxieties "engendered by a progressively precarious world are thereby focused on the city", then Seattle's reputation as, to quote Jonathan Raban, the "city you move to in order to get closer to nature" became a potent symbol of its seeming capacity not only to avoid ecological disaster, but also to promise an assuredness, certainty, and an authentic 'groundedness' within the context of a fragmented and unstable world.[39] As the next section suggests, one of the key ways in which this was articulated in the 1990s was through the enunciation of a resurgent Pacific Northwest regionalism.

SEATTLE AND THE RESURGENCE OF REGIONAL IDENTITY

It has been widely maintained that one of the paradoxes of a global cultural economy is the tendency towards the increasing homogenisation and the increasing heterogenisation of place.[40] Sharon Zukin provides an exemplary instance of this argument when she suggests that:

> The spread of national and even global cultures tends to weaken local distinct-
> iveness … At the same time places have become more differentiated. New

patterns of regional specialisation reflect the selective location of highly skilled and highly valued economic activities.[41]

Zukin's argument about the renewed emphasis upon the importance of enunciating sub-national modes of distinction finds support in what Kevin Robins calls "the new enthusiasm for locality and regionalism".[42] Robins views this trend as a reaction to the universalising forces of modernism, and the desire to seek instead the "continuity in a given place between past and present".[43] Such processes are apparent in the history of American regionalism. As David Wrobel and Michael Steiner point out, American culture resonates with narratives of the death of regionalism. They state, "as the story goes, American regions have been battered by successive waves of nationalism, metropolitanism, capitalism, commercialism, and cyberspace … The death-of–the-region scenario seems embedded in the modern consciousness".[44] However, they contend that American regionalism has in actual fact an "intricate and episodic history"; although notions of the region have served most often as a "largely unconscious source of identity", at certain points it has become a *self-conscious* concern – a cause and a rallying cry.[45] In particular, they note that "end-of-the-century globalism engenders a countertide of pluralism and ethnoregional passions … at the same time that the global village and the World Wide Web promise exhilarating cosmopolitanism, they also stir up longings for more intimate loyalties".[46]

Nowhere has this trend been in evidence more clearly than in the episodic history of regionalism in the Pacific Northwest. Regional identity in the Pacific Northwest "has tended to be somewhat dubious, artificial, and ever-shifting", argues John M. Findlay in "A Fishy Propostion: Regional Identity in the Pacific Northwest".[47] Findlay notes that "until recently, the peoples who have lived in Oregon, Idaho and Washington have generally neither defined themselves readily as belonging to the same region nor, naturally, agreed on the meaning of that region".[48] He adds that "for most of the nineteenth and twentieth century, region has generally fallen behind (as well as in between) the more primary attachments of nation and locality".[49] In addition, as the earlier discussion of the conflict over the region's "natural resources" illustrated, economic development has generated "over time still another set of divisions between metropolis and hinterland".[50] It should also not be forgotten that Washington State is itself characterised by a substantial topographic, cultural and economic division between the areas to the east and west of the Cascade mountain range.[51] Findlay concludes that "as a source of common identity … regional consciousness has seldom papered over effectively the many fissures dividing cultures and societies in Washington, Oregon and Idaho".[52]

The 'Fishy' title of Findlay's study refers to the fact that recent attempts to delineate a "clearly and autonomously defined Pacific Northwest" have fixed upon the salmon as a symbol of an 'essential' and 'timeless' Northwest identity. The attraction of the salmon lay overwhelmingly in its ability to offer an ostensibly 'natural' signifier of identity – what could be more indigenous in the Natural Northwest than the salmon? Yet as Findlay points out, such symbology is somewhat problematised by the fact that the salmon stray beyond the boundaries of even the broadest definition of the Pacific Northwest, to Vladivostok, Siberia and Chongjun, North Korea to be exact. Perhaps more crucially, despite such assertions of the salmon's essential and timeless Northwest identity, Findlay suggests that in actuality the "significance of salmon as cultural 'repository' has changed dramatically over the decades", as it transformed from a disliked dish of local Native American consumption, to a lucrative exported canned good, to a protected species in its supposedly 'natural' state.[53]

Findlay's chronology of the salmon's connotative transformation is useful, but it perhaps underestimates the extent to which the packaging of salmon for consumption remained a key element of the Pacific Northwest's regional distinctiveness throughout the 1990s, and existed somewhat in tension with conservationist formulations of the fish's significance. Indeed, one of the key ways in which articulations of Pacific Northwest regionalism asserted themselves during this period was within the promotional and marketing strategies of commodity culture. Such strategies glossed the inter-regional differences outlined by Findlay, and instead worked to construct notions of "Northwest Lifestyle", closely associated with images of a 'timeless' 'natural' landscape. As David Bell and Gill Valentine note, as regions faced "an increasingly globalising world … questions of regional uniqueness [were] thus distilled into the iconic products of particular places".[54] Moreover, as they make clear, there is no reason why such articulations of regionalism must necessarily originate from the region itself. As Wrobel and Steiner assert, American regionalist discourse has historically been characterised by what they call "interior and exterior regionalism", and it is "the interplay between these two forces" that work to generate notions of the identities of particular regions.[55] Nowhere has that interplay been more apparent than in the essentialised vision of the Pacific Northwest, aimed in particular at upmarket urban consumers, and found in the 'lifestyle' pages of national periodicals.

NORTHWEST LIFESTYLE

Throughout the early 1990s a range of upscale national periodicals published articles that sought to foreground aspects of a 'Northwest lifestyle', valorising the region, and orienting themselves around Seattle. For example, during the period readers could learn about Seattle's Northwest cuisine in the *New York Times Magazine*; the dazzling "Northwest Coast Native American arts" in *Forbes*; the pleasures of living in "A Houseboat in Seattle" in *Esquire*; and Seattle as the epicentre for "voluntary simplicity and spiritual fulfilment" in *U.S. News & World Report*.[56] Whilst Wrobel and Steiner have, with some validity, rebuked such instances of commodity culture for being "regional kitsch for bored urbanites craving tokens of authenticity", this characterisation underplays the significance of these artefacts and their accompanying meanings, in particular the ways in which rhetorics of Northwest lifestyle evoked broader articulations of regionalist discourse.[57]

If one article could be said to encapsulate many of the different aspects that made up the various strands of Northwest lifestyle being promulgated within these periodicals it was a lengthy article written by Ann Japenga, entitled "On a Northwest Course", and published by the *Los Angeles Times* in December 1992. A newspaper with a daily circulation of one million, the *Los Angeles Times* represented a not insubstantial organ of information dissemination concerning Seattle and the Pacific Northwest.[58] Yet it is the way that the article pulled together all the various elements of Northwest 'lifestyle' that make it an exemplary text for analysis.

Japenga's article declared that it was time to "forget adobes and coyotes and deserts [and to] think cabins and salmon and forests, as the anti-style of Seattle takes over what we see, hear and wear".[59] She wrote that:

These days, the style and look reign supreme from a once-neglected corner of the continent … An awesome variety of Northwest goods and services are being packaged and exported … The Eddie Bauer Home Collection offers a riot of north-woods items – faux deer-antler picture frames, trout door knockers and rustic "Northern Exposure" – style furniture.[60]

Japenga noted that "staples of Northwest dining are showing up on menus across the United States … among the most popular food exports are wild mushrooms, wild salmon and wild salad greens".[61] The article's invoking of a 'Northwest style' maintained a focus upon regional identity signified through what Neil Smith calls "the pervasive commodification of nature", and this notion of regional uniqueness was distilled into iconic products that also made 'Northwest style' and Seattle synonymous.[62] This procedure was an entirely cogent one – as David Chaney points out, the concept of 'lifestyle' is tied intimately to a distinctly urban sensibility, emerging as it did with the new forms of social identity and practices shaped by the development of consumer culture within urban modernity.[63] This was apparent in the ways that the article constantly shifted between discussions of 'Northwest style' and Seattle without ever registering a distinction, as the former became appended to metropolitan objects and artefacts of taste and consumption.

Interestingly, the article offered its own theoretical perspective on the phenomenon of Northwest style by including a quote from David Stewart, a professor of marketing at USC, who situated the Northwest style trend within a paradigm of "regional fashion cycles [that] are relatively short-lived, lasting four to six years".[64] Stewart added that such regional fashion cycles reflected the fact that "in our country, we tend to identify with the region where we live and think of others as exotic … now it's the Northwest".[65] Yet this notion of perpetual regional fashion cycles also raised more questions than it answered. It is debatable as to whether all American regions have been incorporated within such 'fashion cycles'. For example, there has been little clamour as yet for "Midwest style", suggesting that the rhetorics of regional valorisation might have a more complex relationship to wider discursive formations than the notion of a hermetic cyclical process would suggest.

Rather than reflecting a self-evident truth, the article used the professor's words to naturalise a notion of regional fashion cycles that it and the retailers quoted were understandably eager to promulgate. The article's situatedness within a newspaper dependant on advertisers' revenue went some way to explaining its generic identity as part story/part marketing patter. For example, the article was the cover story for the *Los Angeles Times*'s 'lifestyle section', a section that also contained adverts for a number of the retail outlets happy to be mentioned, and in some cases provide named spokespersons for the profile of Northwest style. Thus the story made a judicious selection of "Northwest style" signifiers. For example, the article eagerly co-opted a plaid grunge-style flannel shirt look as part of the Northwest's casual "anti-style" but dispensed with grunge music's less marketable associations (for this publication, at least) with heroin use, body-piercing and suicide.[66] In this way, the article needs to be understood as having an investment in the notion of Northwest style that moved beyond the mere generation of story, and into the various strategies of the marketing and promotion of regional identity.

Crucial to the upmarket periodicals that detailed aspects of Northwest lifestyle, and central to Japenga's article for the *Los Angeles Times*, is the concept of 'quality of life'. What was being valorised in the consumables and consumption practices that were outlined was the opportunity to experience the vicarious pleasure of Northwest 'living', which involved the desirable engrossing in a more 'natural' environment. In Neil Smith's terms, what was being sold here is "designer nature" – commodities that "promise urban middle-class consumers a recuperative re-immersion in nature".[67] Indeed, as Ellen Posner points out, not entirely dissimilar concerns with 'quality of life' have been central to many recent academic and popular discussions of urban environmental issues. Moreover, as she notes, what has not often been acknowledged is the fact that, when it comes to urban environments, "quality of life is a middle-class notion".[68] For example, Andrew Ross, discussing the mainstream environmental movement, makes a similar point in highlighting the fact that middle-class constructions of environmentalism "fail to understand how and why the values and activities close to the heart of the population that goes hiking, camping and bird-watching and has easy access to outdoor recreation, simply do not rate too highly among [those] whose access to income and employment and health is persistently endangered".[69] It was precisely this connection between upmarket demographics, the valorising of recreational nature, the promulgation of 'Northwest lifestyle' and notions of 'quality' that migrated across representational forms and into the genres of fictional television programming in the 1990s.

QUALITY OF LIFE TV:
TWIN PEAKS AND NORTHERN EXPOSURE

No discussion of representations of the 'natural' Pacific Northwest in 1990s American culture would be complete without reference to two of the most critically acclaimed television shows of the period, namely *Twin Peaks* (1990–91) and *Northern Exposure* (1990–96).[70] *Twin Peaks*, the creation of film auteur David Lynch and former *Hill Street Blues* writer-producer Mark Frost, had an impact which belied its brief broadcast life, and the show has frequently been talked about in terms of its effect on "changing the face of television".[71] Set in the small town of Twin Peaks, Washington, the show began with the discovery of the dead body of teenager Laura Palmer, wrapped in plastic. The discovery instigated the arrival in Twin Peaks of FBI agent Dale Cooper, who spent the thirty episodes of the show attempting to solve the mystery of Laura Palmer's murder. The show was characterised by a mixture of surrealism, absurdity, formal self-consciousness, intertextuality, and a thematic obsession with extreme violence, corporeality and sexual perversity that also marked Lynch's cinematic oeuvre, and which made its initial appearance on prime-time US network television all the more startling. The two-hour pilot for *Twin Peaks*, aired by ABC in April 1990, was the highest rated TV movie of the season. Lynch and the show became the doyen of TV critics and reviews, and *Time* magazine pronounced it "the most hauntingly original work ever done for American TV".[72] A whole plethora of merchandising tie-ins accompanied the commencement of the show's first series in the spring of 1990, making it a bona fide broadcasting phenomenon. However, the hype did not last long, and as viewing figures began to tail off quickly towards the end of 1990, repeated demotions in ABC's scheduling roster only exacerbated already haemorrhaging ratings until the show was pulled in February 1991.

Whilst numerous accounts of *Twin Peaks* attest to the critical interest in the show from the perspective of formal complexity and televisual postmodern aesthetics, my concern with its significance relates more to the industrial and cultural context of its broadcasting.[73] The movement of an art-house

Figure 5: Nightmares in the 'natural Northwest': David Lynch's *Twin Peaks*

auteur such as David Lynch to prime-time television speaks to a particular moment in American network programming policy. As Robert J. Thompson points out, Lynch's arrival on prime-time has to be understood as part of the ABC network's strategy to compete with rival network NBC in the 'quality revolution,' – itself an attempt to stave off the intense competition for viewers from cable television.[74] As Thompson points out, from an aesthetic point of view, 'quality television' meant, in broad terms, innovative, complex and sophisticated shows, with often controversial subject matter, and associated with a strong writer/creator figure (Steve Bochco being the archetypal example).[75] Indeed, as a number of commentators have noted, many of the precedents for 'quality television' were set up by MTM enterprises in the 1970s, and explored by Jane Feuer *et al.* in *MTM: Quality Television*.[76] Whilst the generic parameters of 'quality television' were necessarily rather fluid, the desired audience demographic could be securely delineated. As Thompson notes, 'quality television' was used by the networks as a means of attracting "an audience with blue chip demographics … upscale, well-educated, urban-dwelling, young viewers advertisers so desire to reach".[77]

Indeed, in the figure of Agent Dale Cooper *Twin Peaks* provided the viewer with the surrogate young urbanite at large in the Pacific Northwest. Although Cooper's gradual descent into madness from an idiosyncratic original self did not allow the viewer a conventional point of identification within the diegesis, he did provide a window of sorts onto the fetishised world of Northwest flora and fauna. *Twin Peaks* dripped with Douglas firs, flannel clothing, native art from the coastal tribes, tastefully decorated lodge-style dwellings, all offering upscale urban viewers a showcase of "designer nature".[78] Each of its immaculately and elegantly detailed episodes basked in the iconic objects used to embody regional uniqueness, and intimately tied to the 'natural' signifiers of place. However, it is important to recognise that *Twin Peaks* was by no means engaged in a social-realist depiction of the region. On the contrary, the show mobilised Northwest nature in service of the creation of a sinister and uncanny milieu – what Brian Jarvis has termed "Northwest Noir".[79] As he notes, the opening credits for *Twin Peaks* serve to generate an image of the natural Northwest as photographed for "a Sears and Roebuck catalogue", whilst the show itself morphed back and forth from pastoral nature to primeval forest.[80] *Twin Peaks* generated a fantastical universe that was at the same time – often in the same frame – a eulogy to an extant Pacific Northwest displaying signifiers of regional uniqueness and charm, and a palimpsest of intertextual references to pre-existing films, novels, folkloric narratives and much more besides. Indeed, part of the pleasure for the knowledgeable viewer was in recognising supposedly significant motifs, homage and references. Yet if Lynch's vision of a diabolical environ was intended to unhinge the tasteful signifiers of 'recuperative nature' abounding in the *mise-en-scène*, it is perhaps apposite that a show that frustrated all attempts to meld allusions and scattered intertextual fragments into a cohesive metanarrative saw itself pilfered for some of its very tasteful signifiers, this time harnessed to a much gentler vision of the natural northwest, in the shape of the comedy drama *Northern Exposure*.

Like *Twin Peaks*, *Northern Exposure* organised its initial storyline around the entry of an urban East-coast inhabitant to the Pacific Northwest, this time in the guise of New York medic Joel Fleishman. *Northern Exposure* was set in the fictional town of Cicely, Alaska, although filmed in Roslyn, eighty miles east of Seattle. Moreover, the "real" geographic identity of the show was hardly a secret, a fact demonstrated by the brisk tourist activity experienced in Roslyn during and subsequent to its broadcast.[81] Running for over one hundred episodes and six years, *Northern Exposure* was a sustained ratings success, particularly in its first three seasons, and also garnered high critical praise, winning numerous awards, including an Emmy for best dramatic series, and two Golden Globes.[82]

Northern Exposure retained the off-beat storylines and eccentric characters of *Twin Peaks*, and depended upon a similarly extended ensemble cast to create multiple storylines. Reflecting its generic identity as a comedy drama, the show generated a more accessible and user-friendly metaverse of nuanced characters and whimsical plotlines – to use Thompson's adjective – "quirky" rather than diabolic. As he notes, the cornerstone of the show was spirituality – to be precise, a concern with 'new age' spirituality that for the most part rejected organised religion in favour of a brand of pre-modern spiritualism. For example, 'Aurora Borealis', the episode that ended season one of the show, provided a not untypical example of such spiritualism, and depicted the characters preparing for the approach of the Northern Lights. The plot of the episode revolved around the strange behaviour exhibited by the characters as they responded to the impending natural phenomenon, for example rejecting their quotidian activities in favour of actions and deeds that displayed a primordial connection to the rhythms of Gaia. Rejoicing in a harmonious relationship with the Pacific Northwest's wondrous surroundings, the characters' interaction with nature tapped into narratives of postmodern environmentalism, and what Neil Smith terms a "frothy ideology of recuperative nature" mobilised so lucratively by high-end retailers towards receptive urban consumers in the 1990s.[83] Indeed, what *Northern Exposure* shared with *Twin Peaks* was a raft of commercial tie-ins and merchandising, including CDs, photo albums and *The Northern*

Exposure Cookbook, an explicit attempt to cash-in on the vogue for Northwest cuisine in the 1990s, and to capitalise on the wealthy blue-chip demographic for such instances of 'quality television'.

There is, obviously, much more that could be said about both shows, and my brief exegesis necessarily does a disservice to their respective narrative and visual complexities.[84] However, my intention here is to highlight a particular range of social and cultural determinants at work in generating the context for the production and consumption of the shows. That both *Twin Peaks* and *Northern Exposure*, devised independently, and at the same time, mobilised narratives of 'Northwest lifestyle', and valorised recreational nature for upscale urban viewers at precisely this moment points to the need to understand them not only as the creation of autonomous televisual auteurs, but also in relationship to other aspects of American popular culture in the 1990s.

NATURALLY INSPIRED BY SEATTLE: EDDIE BAUER AND REI

Recognising the various functions performed by representations of nature in American popular culture can extend beyond a consideration of cinematic, televisual, novelistic and journalistic accounts. No assessment of Seattle's 'natural' identity in the 1990s would be complete without a consideration of the role played by the city's two leading retailers of outdoor apparel, namely Eddie Bauer Inc. and REI (Recreational Equipment Inc.). An examination of the respective retailers' merchandising strategies and promotional materials illustrates the ways in which they functioned as complex textual environments that worked to generate profits from mobilising discourses, symbols, metaphors and fantasies of 'nature'.

Eddie Bauer Inc. and REI were both founded in Seattle, and by the end of the 1990s still retained their main headquarters in the city. Crucially, the period also saw them both draw significantly upon their Seattle origins as part of the marketing strategies for their brands, and they sought actively to associate their product ranges with the ethos of 'Northwest lifestyle' outlined in this chapter. For instance, Eddie Bauer's advertising literature described the aim of its commercial operation as the merchandising of "distinctive clothing, accessories and home furnishings for today's active, casual lifestyle" through

So close,

Figure 6: Nature's playground: kayaking in Seattle's Elliot Bay

its two retailing concepts: Eddie Bauer® and Eddie Bauer Home.[85] By the end of the decade the company had over 600 stores in the US, Canada, Japan, Germany and the United Kingdom, based mainly in urban and suburban malls and shopping complexes, while also retailing online and through its mail order catalogue, and in 1999 registered $1.7 billion in sales.[86] Eddie Bauer's internal marketing mantra "think Seattle, act globally" reflected the fact that the company marketed itself aggressively in connection with the city of Seattle. For example, Eddie Bauer's movement into the United Kingdom in 1996 was accompanied by glossy adverts in the lifestyle sections of broadsheet newspapers such as the *Observer* and the *Sunday Times*, the centrepiece of which was a large photograph of the Space Needle. The advertising campaign also included a competition to win an all expenses paid trip to Seattle, thus reinforcing the close association it wished potential customers to make between the brand and the Pacific Northwest city. Eddie Bauer's more routine range of advertising material was also been careful to foreground its Seattle 'heritage'. Store leaflets, mail order catalogues and the Eddie Bauer website all included its story of origin, stating that "since 1920, Eddie Bauer has developed casual, comfortable, and stylish apparel and gear for the quality conscious consumer … inspired by the natural beauty and ruggedness of the Pacific Northwest", and include the fact that the first store was established in downtown Seattle in 1920.[87]

Eddie Bauer's counterpart REI described itself in promotional literature as "the leading international retailer of quality outdoor gear and clothing".[88] Like Eddie Bauer, REI sold a wide range of clothing, and also stocked climbing, hiking, fishing and other outdoor sports equipment. Although by the end of the 1990s REI had only 54 retail stores in 23 states and in Japan, which made it relatively small in comparison with Eddie Bauer, it registered sales of $621 million in 1999, aided by its long-established mail order business. As the nation's largest consumer co-op REI had more than 1.5 million active members by the end of 1998, although co-operative membership, purchasable for $15, amounted to little more in practice than a "dividend" based on each member's annual purchases, making it akin to a host of store loyalty cards.[89] REI's advertising material similarly foregrounded its Seattle heritage, for instance stating that the company "started in 1938 when Seattle mountaineer Lloyd Anderson sent away for a new ice axe".[90] This reference by both companies to pre-World War Two origins was far from superfluous – not only did it serve to associate their brands with "tradition, quality, and craftsmanship", but also to imbue a vague "authenticity" through their connection to rugged "mountain men" founders. Indeed, both Lloyd Anderson and Eddie Bauer featured in "founder narratives" on the companies' web sites and literature: Anderson depicted at the summit of Jungfrau, Switzerland in 1959; Bauer referred to as a "legend", with homage paid to his adeptness as a hunter and fisherman (safely located in another supposedly less resource-scarce era).[91] As David Bell and Gill Valentine point out, this invoking of 'tradition' is a familiar trope of advertising campaigns for brands and commodities trading on notions of 'authentic' regional identity, whilst simultaneously operating within global consumer culture.[92]

Although both Eddie Bauer and REI emphasised their lengthy heritage in promotional material, a close inspection of their evolutionary timelines indicates that they both experienced a quite recent and rapid expansion. Founded in 1938, REI's first full-time retail store opened in 1944 in Seattle. It did not open its second store until 1975, in Berkeley, California, and as recently as 1983, REI had only nine retail outlets. From 1988 onwards, REI began to open three to four stores per year.[93] Similarly, Eddie Bauer Inc., although founded in 1920, with Eddie Bauer's Sports Shop in downtown Seattle, did not open its second store (in San Francisco) until 1968.[94] As late as 1988, the company had only 61 stores in total. It is clear that the expansion of the two companies coincided with the rapid rise in participation in outdoor recreation activities since the 1980s, a fact outlined in the literature of the Outdoor Recreation Coalition of America (ORCA).[95] ORCA's 1994 survey of trends and developments in outdoors recreation concluded that "for most [outdoor] activities, there are millions more participants in 1994 than there were in 1983".[96] Interestingly, the survey also concluded "the overwhelming majority of enthusiasts … can be considered upper-middle to middle class" and that "this is an important consideration for the marketing of outdoor recreation equipment and services".[97]

However, it is also the case that the rapid expansion of both companies in the 1990s accompanied Seattle's unprecedented and considerable rise to prominence. That both companies saw such rapid expansion in the 1990s attests in part to the astuteness and perspicacity of harnessing their brand identities to the city and its natural appeal. Indeed, neither company failed to recognise the potential profits to be made from aligning themselves with other lifestyle commodities that traded heavily and aggressively upon an association with Seattle. In October 1997, Eddie Bauer initiated a joint venture with the Seattle gourmet coffee company Seattle's Best Coffee (SBC). Seattle's Best Coffee launched what it called its "Eddie Bauer Blend" coffee. The company's advertising material, carried by national newspapers to coincide with the launch, stated that "there is a natural fit between Eddie Bauer and Seattle's Best Coffee. It starts with our common Northwest heritage, and encompasses our shared values and goals surrounding quality and unmatched customer service".[98] Along similar lines in March 2000 REI announced that its "flagship" stores in Denver and Tokyo would incorporate Starbucks Coffee shops. The company's press release stated that "coffee stores are very popular with our customers and enhance the experience of REI being a clubhouse for outdoors people".[99] As chapter six points out, gourmet coffee has had its own instrumental role in the signification of the city in the 1990s. Here, it is enough to note that the enunciation of "common Northwest heritage" could extend to other brands and products trading on notions of 'authentic' regional identity.

This chapter has has historicised and contextualised one particularly important aspect of Seattle's relationship to its natural surroundings, namely the notion of the city at the heart of a 'natural northwest' in the 1990s. What is clear is that during this period the notion of Seattle as the 'city you move to in order to get closer to nature' was tied intimately not only to the idea of nature as an 'urban appendage', but also to the promotional and marketing strategies of commodity culture, which mobilised images and narratives of the 'natural northwest' for middle-class urban consumers across the United States and beyond. The potential appeal of motifs of nature in connoting 'authenticity', 'quality', 'tradition', 'heritage' and the enunciation of northwest regional uniqueness can be explained not only in terms of the attractive contrast with other urban areas suffering from a myriad symptoms of environmental degradation, but also the suggestion of an assured and grounded natural 'sense of place' within the context of a fragmented and unstable world.

Although the paeans to "common Northwest heritage" and "outdoors people" utilised by Eddie Bauer and REI were part of the companies' lightweight promotional patter – the seemingly vacuous and breezy signs offering little of significatory substance, it is worth pointing out that articulations of regional identity have manifested themselves in more problematic defensive formulations in recent times.[100] In other words, the retrograde rhetorics of "regional pride" have been invoked as a secure and stable affirmation of identity within a period of rapid change, and with cultures seemingly changing and mutating within a globalising world. Such turns to homogenous cultures, "the contiguous transmission of historical traditions or 'organic' ethnic communities" has been associated with much more troubling pronouncements concerning regional essentialism.[101] Moreover, it is also the case that the "heritage" of such profiling of Northwest identity does have a rather unsavoury historical precedent. As Richard Dyer notes, "the Aryan and Caucasian model share a notion of origins in mountains". Noting also the Romantics' penchant for "small, virtuous and 'pure' communities in remote and cold places" he adds that:

> Such places had a number of virtues: the clarity and cleanliness of the air, the vigour demanded by the cold, the enterprise required by the harshness of the terrain and climate, the sublime, soul-elevating beauty of mountain vistas, even the greater nearness to God above and the presence of the whitest thing on earth, snow. All these virtues could be seen to have formed the white character, its energy, enterprise, discipline and spiritual elevation, and even the white body, its hardness and tautness.[102]

It is with such delimited racial profiles of the Pacific Northwest in mind that the next chapter seeks to gain some purchase on more extensive narratives of Seattle's racial identity in the 1990s.

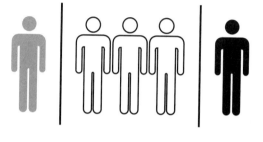

CHAPTER 3

DREAMING OF A WHITE SEATTLE

"I don't think I've ever met a brother from Seattle in my life, man…"[1]

The fourth season of the Emmy-award winning HBO comedy *The Larry Sanders Show* (1992–98) included an episode entitled "Larry's Sitcom". Like most episodes of this 'behind the scenes' fictional spoof of a prime-time network talk show, the events of "Larry's Sitcom" revolved around the various trials and tribulations that comprised the relationship between the talk show's assorted creative personnel and the 'suits' of the network. Indeed, the show used its identity as a cable production to satirise mercilessly the machinations of network executives, and their relentless desire to prize audience ratings over and above creativity, artistry and innovation. In "Larry's Sitcom" the central dilemma related to the difficulties encountered by the network as they formulated ideas for a new sitcom. The sitcom was being developed as a star vehicle for the black actor David Chapelle (playing himself in the episode). A draft script had been written, laying out the key elements, including the intention to set the show in Seattle. At a network meeting for script development, Chapelle rejected the proposed location for the sitcom, pointing out that he had never "met a brother from Seattle in my life, man", and added for emphasis the opinion that Seattle was not a "place where [black people] would really, really live". After assuring all involved, in typical network-speak, that they "really loved the show", the network executives departed. The episode ended with the knowledge that the sitcom, starring David Chapelle, had indeed been commissioned, but had been relocated to Baltimore.

Although *The Larry Sanders Show* was fictional entertainment, much of the appeal and power of the show stemmed from its verisimilitude, in particular the way that it mobilised widely circulating knowledge of the entertainment industry, and its relationship to the wider currents of American society, in the service of innovative generic material. Thus the humour of "Larry's Sitcom" relied on the fact that David Chapelle's rejection of a sitcom starring a black actor and set in Seattle would be plausible for the viewers, reflecting as it would their own sense of the credible parameters for a Seattle-set network sitcom. As this chapter's discussion of *The Larry Sanders Show*'s Emmy awards rival *Frasier* will go on to illustrate, the instincts of Chapelle and the fictional network executives were unerringly correct. Indeed, a perusal of actual American network programming in the 1990s indicates that there was more than a little substance to Chapelle's claim. The evidence would seem to suggest that while shows with prominent black characters did take place in New York (e.g. *Living Single*, *NY Undercover*, *NYPD Blue*), Los Angeles (e.g. *The Fresh Prince of Belle Air*, *The Show*, *South Central*) or, following "Larry's Sitcom", Baltimore (e.g. *Roc*, *Homicide: Life on the Street*), they did not take place in Seattle.[2]

Yet Chapelle's suggestion that Seattle was not a "place where [black people] would really, really live" was clearly not true. By the late 1990s Seattle's black inhabitants made up ten per cent of the metropolitan populace, roughly proportionate with their population percentage in the United States as a whole.[3] However, this proportion was relatively small for an American city, making Seattle "not even in the top 50 cities in terms of [the black] percentage of the population".[4] Furthermore, Seattle's ten per cent was a relatively recent development: pre-World War Two Seattle's black population ran at a steady one per cent, rising to three and a half per cent by 1960. Such statistics could be used to contend that in comparison with, for example, Chicago, New York or Atlanta, Seattle had not functioned, historically, as a substantial centre for a black urban populace. It had therefore not functioned as a significant location for the generation of black urban cultural forms, one of the main historic catalysts for representation, as apparent from examples such as New Orleans jazz, or the Harlem Renaissance, or more recently, New York hip hop. On the contrary, two of Seattle's few prominent and influential black artists, namely Quincy Jones and Jimi Hendrix, both left the city in order to establish their careers, a fact that perhaps reinforced the impression that the city lacked the artistic community to support and sustain black talent.[5]

A similar claim could be made for Seattle's Asian-American community, home at one stage to two of the most celebrated Asian-American actors of all time, namely Bruce Lee and Keye Luke. [6] Seattle had a Japanese population since the 1890s, and a Chinese population earlier still. However, its growth in becoming, at 11.9 per cent, the city's largest minority, was a relatively recent phenomenon, stimulated primarily by a significant influx of arrivals from Cambodia, the Philippines, Samoa, Thailand and Vietnam in the 1990s, a trend which resulted in the number of Asian American inhabitants increasing by 48 per cent between 1990 and 1996.[7]

Yet it is worth highlighting the fact that the historically small numbers of minorities in Seattle cannot be disassociated from the prevailing attitudes and actions of generations of white settlers in the Pacific Northwest. Indeed, it is not too far-fetched to submit that many such settlers were engaged in persistent attempts to achieve the actual, physical effacement of the entire range of the region's non-white population.[8] As John M. Findlay points out in "Regional Identity in the Pacific Northwest":

> Over the years … whites in the Pacific Northwest have discouraged African-Americans from coming; coerced Indians onto reservations and away from economic opportunities; lobbied the federal government to restrict or halt Asian-American immigration; pressured Chinese immigrants to leave, sometimes violently; prevented Japanese immigrants from owning land; and supported both the internment of people of Japanese descent during World War Two and their continued exile from the region after the war.[9]

As Findlay indicates, the Pacific Northwest has therefore been predominantly white "by both circumstance and by design", and that despite a substantial change in the composition of the region's population since World War Two, it should "come as no surprise … that American white supremacist groups in the late twentieth century have been attracted to the region as one place where their goal of an exclusively white population seems relatively attainable".[10] Certainly, the rhetoric of 'regional pride' that bolstered visions of 'northwest heritage' and 'outdoors people' enunciated in the promotional material of

Eddie Bauer and REI discussed in the last chapter intimated an inclination on the part of some to cultivate a homogenous and cohesive collective identity for the city and for the region. It is also the case that the range of national periodicals examined in chapter one similarly reinforced an impression of the city's homogenous racial profile, from *Lear's* portrait of Seattleites as "typically tall and fair, hearty and adventurous pioneers" to the *Christian Science Monitor's* veneration of "an unpretentious city modulated by a polite, if cool, Scandinavian clannishness".[11]

It is important to note that a number of Seattle's novelists published work in the 1990s that made crucial steps to narrativise both the difficult history and the shifting identity of the city's racial demographic.[12] For example, the Filipino-American writer Peter Bacho generated a fictionalised history of Seattle's Filipino community in his debut novel *Cebu* (1992), which won the American Book Award in 1992. The novel told the story of Ben Lucero, a middle-aged Filipino-American from Seattle, who returned to his mothers' hometown of Cebu, near Manila, for her funeral. The visit, and the many encounters it engendered, propelled Lucero to question his faith and many of his beliefs about notions of belonging and origins upon his return to Seattle. A key theme in the novel was the profound ambivalence felt by many first-generation Filipino-Americans regarding their parents' emigration to America, in this case triggered by Lucero's return to Cebu. Indeed, Bacho has not been alone in his attempts to trace the difficulties of immigrant life in Seattle. Lydia Minatoya's debut novel *The Strangeness of Beauty* (1999) created a powerful work of fiction around the harsh lives of Japanese-Americans in Seattle in the 1920s. The novel was constructed as the mock-autobiography of Etsuko Sone, who leaves Kobe, Japan in 1921 to move to Seattle. Minatoya sets up Etsuko's preconceptions of Seattle as a west-coast New York, bustling with the kineticism of the modern American metropolis, full of temples of learning and vibrant, prospering immigrant communities. She then charts the reality of Etsuko's arrival in a frontier town still comprised in large part from the surrounding soil and forests. The story was a bleak tale of dashed expectations, as Etsuko's husband and sister both die in Seattle, the first from a fishing accident, the second from childbirth. The sense of failure and despondency accompanying the schism between the promise and the reality of the immigrant experience is explored in detail, particularly in the later sections of the novel that chart Estuko's difficult return to Kobe.

The purpose of briefly delineating these two novels is not to set them up as particularly representative. Rather, it is merely to point out that Seattle has not been exempt from the forces of migration and what Stuart Hall terms the "proliferation of new identity positions" bequeathed by the new diasporas of late modernity, and the multiform attempts to enunciate those experiences through the cultural work of representation.[13] As Homi Bhabha points out:

> The historical and cultural experience of the western metropolis cannot now be fictionalised without the marginal, oblique gaze of its postcolonial, migrant populations cutting across the imaginative geography of territory, community, tradition and culture.[14]

Yet despite the fact that Seattle has also been the object of the marginal, oblique gaze of its postcolonial, migrant populations, and that such works of fiction have been critically well received – as well as winning an American Book award, *Cebu* has found its way onto a number of Asian-American literature courses in US universities – it is also true that they have not been influential in shaping mainstream understandings of Seattle.[15] In other words, despite the fact that Seattle has always been a multiracial and multicultural city, its novelists have not impacted upon the popular imaginary of the city in the way that, for example, African-American literature has shaped understandings of New York or Chicago, or the impact of black and Asian writers such as Salman Rushdie, Hanif Kureshi and Zadie Smith on the imaginative geography of contemporary London. On the contrary, when it came to creating an influential and enduring imaginary landscape of Seattle in the 1990s, it seemed than any colour would do, so long as it was white.

FROM LARRY SANDERS TO THE L.A. RIOTS

The relationship between profound cultural, economic and socio-spatial transformations and the significant increase in non-white populations has been one of the defining features of the recent history of America's cities. As Stuart Hall notes, the "continuous large-scale, legal and 'illegal' migrations into the

US from many poor countries of Latin America, and the Caribbean basin" as well as significant numbers of "'economic migrants' and political refugees from South-East Asia and the Far East" has transformed fundamentally the ethnic profile of many of the nation's cities.[16] Nowhere has the confluence of patterns and forces resulting in what Hall terms "'The Rest' in 'The West'" manifested itself more explicitly in material terms than in the tendency of large urban areas to ghettoisation. The processes of the advanced global political economy have been reflected in what James Donald terms the "complex texture of urban space", as the selective restructuring of capitalism has been instrumental in the creation of blighted urban communities marked by structural unemployment, homelessness, poverty and criminality.[17] Such developments reflect what Doreen Massey calls the "power-geometry" of time-space compression, and the fact that "some are more on the receiving end of it than others; some are effectively imprisoned by it".[18] Echoing such carceral imagery, Mike Davis refers to the "hardening of the urban surface in the wake of the Reagan era", and maps out the "brutalisation of inner-city neighborhoods" and the apartheid spatial relations of contemporary Los Angeles.[19] The bifurcation of patterns of uneven development along racial lines thus finds its corollary in the symbolism of the built environment, and what Sharon Zukin terms the "aestheticisation of fear" apparent in privatised public spaces, technologies of surveillance, and the secure upscale residential enclaves of New York.[20] As Zukin notes, in this urban landscape of fear "there are no safe places. The Los Angeles uprising of 1992 showed that, unlike in earlier riots, the powerless respect fewer geographical boundaries."[21]

Zukin's notion of an 'aestheticisation of fear' draws heavily on the idea of 'white anxiety' as a tangible response to profound urban socio-spatial restructuring, and one that has also found its way into the cultural work of representation. As Donald notes, one response to the 'flux of community and economic globalisation' has been provided by images in the mass media, which, he argues, have become "increasingly *racialised*".[22] A crucial vehicle for such racialised imagery is the journalistic accounts of urban decline examined by Robert Beauregard in *Voices of Decline.* Unpicking the rhetorical tropes mobilised by the discourse as articulated within the mainstream American press, Beauregard explains how the ghetto has served as "the symbolic reference for many social ills … and the city itself", whilst "race is the figure in which is condensed the perplexing problems" of the city.[23] As the discussion of the gentrified city 'frontier' and urban 'pioneers' in chapter one makes clear, any consideration of the myriad profiles of "urban decline" carried by the mass media needs to be sensitive to the ways in which the phrase has frequently functioned as a euphemism for race.[24]

Beauregard points out that the discourse on urban decline first took on a specifically racial dimension during the 1960s, as "racial disorders in Birmingham, Savannah, Cambridge (MA), Chicago and Philadelphia began to reshape how urban decline was being represented". At this time race became the "single issue around which all other symptoms and causes were arranged".[25] However, he maintains that after the upheaval of the 1960s the discourse of decline began to be pushed along other dimensions, most particularly into narrating and shaping responses to the impact of the severe and widespread economic recession of the early 1970s, and which had resulted in the financial collapse of the nation's large industrial cities.[26] What brought "race back into the debate [on urban decline] with a vengeance" were the riots in Los Angeles in April 1992, ignited by the Rodney King "not guilty" verdict. Coverage of the riots in the national press made references to the race riots of the 1960s, and underlined "the volatility of cities as a symbol of American race relations".[27] Although, as Darrell Hamamoto notes, Asian-Americans, so long depicted as America's "model minority", were drawn into the disturbances as a number of clashes between Korean-Americans and African-Americans occurred, Davis points out that "most of the news media remained trapped in the black and white world of 1965".[28]

What is clear is that against the backdrop of national newspaper and television reports of the anger and alienation of Los Angeles, Seattle began to function in the news media as something of an urban oasis for jittery white middle-class professionals. For example, not untypical in this regard was an article by film critic Michael Medved, entitled "Paradise Lost – Riots, Fire, Slump and now Earthquake – for many Californians the dream is over" and carried by the *Los Angeles Times*. The article centred around the purportedly archetypal quote from an "independent businessman and father of five" who stated that "things have been going wrong in this town for a long time … anybody who cares about his children has to ask himself, why should I make my children put up with more of this? We're talking about going up to

Seattle."[29] This depiction of Seattle as a locus for 'white-flight' was echoed in another article by Martin Walker for *Reuters* in September 1992, and carried by the *New York Times*, amongst others. The piece was entitled "Washington State replaces California as haven for infrastructure investment", and Walker wrote that:

> The Californians came here from the Mid-West attracted by good schools and highways. Now they see crime in the streets, riots in Los Angeles, a budget crisis and years of down-sizing ahead … the Californians, deprived of that quality of life which intelligent public spending can provide, should love [Seattle].[30]

The implication is that for such Californians, Seattle represented a "Paradise Found" – an economically healthy urban environment free from crime and rioting (and what such rioting euphemistically symbolised) in which to bring up their children. In such narratives, it was not so much that Seattle's non-white population was being "forgotten", but rather, that Seattle's whiteness was being actively mobilised. This was also given support in an article by the *Los Angeles Times*' writer Joel Kotkin in March 1993. Kotkin disputed the fact that some commentators were suggesting that Californian 'white flight' to Seattle simply signalled the emergence of the city as "a new economic power centres".[31] Instead, Kotkin suggested that:

> Outbound California migration seems more driven by middle-class families, especially white, fleeing the stresses of urban life for calmer, more culturally homogeneous environments. Fear may also be a motivation – fear of a multiracial society in which Latinos, African-Americans and Asians are key players.[32]

As Mike Davis points out in *Ecology of Fear*, white flight from Los Angeles had been occurring in a quiet and orderly fashion long before the Los Angeles Riots, but the events in 1992 exacerbated exponentially the sense of 'white anxiety', as the print press focused upon 'black rage', reflecting an obsession with black violence that seriously under-represented the multicultural nature of the civil unrest, just as it over-represented the racial homogeneity of Seattle as 'white flight' centre.[33] Whilst I am not suggesting that the newspaper accounts of Seattle referred to above were in any sense definitive, neither were they untypical of the way in which the city came to function as an emblematic site within the mainstream press, most particularly against the backdrop of a climate in which, as Liam Kennedy points out, cities became "more intensely psychologised as sites for racial anxieties".[34]

As Kennedy's own work suggests, anxieties pertaining to ethnicity and race worked themselves into representational form in a range of concurrent literature and film that sought to situate stories in America's large urban centres. Within this work, Kennedy pays particular consideration to "the growing visibility of whiteness as a social category" and the increasing body of work within the humanities that has sought to analyse depictions of whiteness.[35] As he notes, the theoretical concern with whiteness as a racial identity (as opposed to interest in white "ethnicities" such as Jewishness) is a quite recent development. Indeed, the publication of Richard Dyer's article "White" in *Screen* in 1988 can be regarded as one of the first attempts to consider the racial imagery of white people in popular culture, and was later expanded into a book-length study of the subject.[36] Dyer's explanation for his examination of whiteness sets out quite clearly what he considered innovative about the mode of analysis:

> Until recently a notable absence … has been the study of images of white people. Indeed, to say that one is interested in race has come to mean that one is interested in any racial imagery other than that of white people. Yet race is not only attributable to people who are not white, nor is imagery of non-white people the only racial imagery.[37]

Dyer points out that the assumption that "white people are just people" is played out in the "invisibility of whiteness as a racial position in white discourse".[38] As Dyer suggests, this is not to say that there is "no discussion of white people. In fact for most of the time white people speak about nothing but white people, it's just that we couch it in terms of people in general."[39] In addition, as John Gabriel notes, one of the "reasons why whiteness keeps itself so well hidden is because it works through other discourses".[40] Dyer asserts the importance of studying whiteness in the face of such white "invisibility", stating that "we won't get to [genuine multiplicity without (white) hegemony] until we see whiteness, see its power, its particularity and limitedness, put it in its place and end its rule".[41] Recent important publications such as bell hooks' *Black Looks* (1992) and Fred Pfeil's *White Guys* (1995) have also engaged in the project of making visible the contemporary images of whiteness – an activity referred to by Dyer as the process of "making whiteness strange".[42]

It is no coincidence that a theoretical interest in whiteness corresponded with an increasing awareness of an "incipient crisis of whiteness" in American and Western European culture in the 1990s.[43] Dyer acknowledges this obliquely whilst expressing concern that writing about white people might result in ostensibly giving "white people the go-ahead to write and talk about what in any case we have always talked about: ourselves".[44] With this in mind he cites a *Newsweek* cover story from 1993 about white male paranoia, and the rise of neo-fascist political parties in Europe and North America as symptoms of a troubling declaration of white identity. In contrast, John Gabriel is more disposed to sketch the parameters of these phenomena as instances of what he terms "uncoded" whiteness. He situates a new "crisis of whiteness" within a "set of new global conditions", elements of which have already been outlined by this study, including global migrations and diasporas; liberation and independence of former colonies; and a more assertive politics of representation "concerned with issues of bias, positive and negative images and access to media institutions" by non-white Americans, and that "the fact that whites are numerically becoming a minority in the US provides an important material dimension to these anxieties".[45] Nowhere has this process been apparent more visibly than in the nation's major cities, which have also been the site of some of its most powerful fictional dramatisations.

Within the critical literature that has sought to examine representations of whiteness in American culture, few narrative fictions have received the attention afforded the Hollywood film *Falling Down* (1993). Set in contemporary Los Angeles, *Falling Down* tells the story of a middle-class white male, known only as D-FENS, who embarks on an impromptu journey across the city after losing his job. Estranged from his wife and child, D-FENS' spiralling sense of helplessness and anger leads him to vent his frustration on the various individuals and groups he meets on his demented picaresque movement across the city, taking it in turns to clash with a Korean shopkeeper, Latino gang members, a neo-Nazi and elderly country club members before being shot by a white policeman of not dissimilar age and background. Such individuals, and the fragmented landscape they inhabit, are sketched as emblematic of the social, cultural and spatial reorganisation of the contemporary city, a city from which D-FENS, as 'downsised' white middle-class male, feels profoundly dislocated. As noted by Richard Dyer, Fred Pfeil, Jude Davies, John Gilbert and Liam Kennedy, the film exhibits a considerable ambivalence towards its central character, and the morality of the position he occupies.[46] Indeed, the overwhelming interest for commentators has been in trying to untangle the confused and conflicting messages the film supplies about white (straight) male identity, and the moral and empirical legitimacy of white urban anxiety. What is not in question is the status of the film as a key reference point for writers seeking to examine representations of whiteness and the contemporary American city, and the ways in which such representations exist in a complex relationship with the social reality they purport to reflect.

Yet *Falling Down* was by no means the only high-profile and commercially successful Hollywood film to offer a depiction of white anxiety in the contemporary city at around that time. Released just a few months previously, the Seattle-set film *The Hand that Rocks the Cradle* had been the biggest box-office success of 1992. Whilst the attention afforded *Falling Down* can be attributed in large part to its (desired) status as 'agenda setting' – that it sought to make an explicit statement about white anxiety in Los Angeles that spoke to the complex race relations in the city, *The Hand that Rocks the Cradle* was received and understood quite differently. Although, like *Falling Down*, it was an adult-oriented film, generically it functioned in contrasting ways. *The Hand that Rocks the Cradle* is a thriller, and one that does not purport to offer an overt statement on white identity or contemporary race relations. Indeed, as the film lacks *Falling Down*'s uninhibited histrionics and explicit social messages, *The Hand that Rocks the Cradle*'s reviewers concentrated on its proficiency as a thriller. Not untypical were the reviews in the *Washington Post* and *Rolling Stone* which called the film "a luridly efficient thriller" and "a cheap thrill and scary fun" respectively, whilst also noting the Seattle setting.[47] What the reviews did not articulate was the fact that the film drew just as heavily upon the prevailing image of its respective city's racial profile as did *Falling Down* in order to generate an equally compelling but quite different vision of white anxiety. If *Falling Down* mobilised pre-existing, racialised visions of Los Angeles as an 'urban jungle', then *The Hand that Rocks the Cradle* drew upon Seattle's pre-existing racialised vision as a liberal 'white flight' oasis, and undermined it with glee and venom.

Set in the Magnolia neighbourhood of Seattle, *The Hand that Rocks the Cradle* told the story of Peyton Flanders, a gynaecologist's widow (played by Rebecca de Mornay) who seeks revenge against Claire Bartell (Annabella Sciorra). Bartell's accusations of sexual assault led Flanders' husband to kill himself, and caused Flanders to lose her own unborn child. Under the guise of a nanny, Peyton works herself into Claire and her husband Michael's home and begins gradually to destroy their family from within.

The first shot of the film fixed the image of the Bartells' house at dawn: large, detached, well kept and above all, painted gleaming white. The rest of the title sequence inter-cut shots of the house interior with a figure pedalling a bicycle laboriously along the road. The interior of the house was also overwhelmingly white; white walls, white doors and white railings. As the sun came up, the house became ever more resplendent in its whiteness. At the same time, repeated shots of the cyclist revealed more of his body – at first a leg, then a torso, until eventually we see the whole figure, except for the head, which is obscured by a hooded top. We then see shots of the white family engaged in routine morning activities. Claire is making juice in the kitchen, Michael is shaving upstairs, watched by their young daughter Emma. We are then shown that the figure that knocks on the door of the big white house is African-American. Understanding that his knocks on the door are receiving no reply, the man walks round the side of the house, and past the kitchen window. After gradually building up the tension, the next shot sees Claire spot the man through the window, scream, drops her jug of juice, and yell for Michael. Michael races downstairs, moves outside, apprehends the man, and asks him "what are you doing here?"

The sequence is quite perfectly set up for symbolic effect: the gradual revelation of the brilliant whiteness of the domestic environment, juxtaposed with the equally gradual revelation of the blackness of the figure that approaches. Both moments collide with Claire's scream, a scream that ruptures the perfect white family morning. Michael's question "what are you doing here?" is addressed to the 'trespasser' individually, but also *racially* – the whole sequence is set up to underscore the dramatic and unexpected entrance of blackness on perfect whiteness.

The movie quickly makes clear that the man is Solomon (Ernie Hudson), who has been sent to the Bartells' house by the 'Better Day Society', an organisation that finds work placements for mentally disabled people. Solomon has come, as requested, to build the Bartell family an appositely symbolic white picket fence. Solomon, seemingly lacking the guile that would deter him from asking, inquires as to whether the fence is intended to "keep people in, or keep people out?" Claire is surprised by the question, but recovers to answer, rather sheepishly, "both – mostly out". Solomon's question is a crucial one, since the whole movie is structured around a set of assumptions underscoring who is and is not allowed into the Bartells' home. Solomon's mental disability serves (at least initially) to diminish the potential "threat" of him as a powerfully built black male, yet the limits of his embrace into the white home are clearly demarcated. He is not allowed to touch or pick-up the Bartells' infant child, despite the reasons for this never being stated – in other words, it is simply "common sense". Thus the Bartells' benevolent liberalism is not allowed to interfere with their better judgement about the safety of their child.

In contrast to Solomon, Peyton has no problems gaining instant access and acceptance by the Bartell family. Unlike Solomon, Peyton does not come to the Bartells' home with impeccable references, yet she does not need to, since she has her own permit to entrance: the palest skin imaginable, complimented by light blond hair. As Richard Dyer points out, "whiteness as a coalition also incites the notion that some whites are whiter than others".[48] An ostensible Snow White, the Aryan Peyton is the palest of them all – in this most symbolic of ways she fits in, and is consequently given complete trust over the Bartells' infant, in her role as the nanny. Peyton proceeds to exploit systematically this position of trust, using her understanding of the Bartells' residual "common sense" fear of Solomon to frame him for inappropriate behaviour towards Emma, until Claire, during a visit to Peyton's own (pure white) former home, realises her nanny's true identity. Peyton is vanquished finally after a struggle by being impaled, fittingly, on the Bartells' white picket fence. In the film's dénouement, and against the backdrop of this broken fence, Solomon is given the ultimate symbol of the Bartells' trust and inclusion, namely the white baby to hold.

The Hand that Rocks the Cradle manipulated for thrills, quite beautifully, the wider currency of Seattle as a 'white oasis'. The Bartells' residual white

Figure 7: Peyton (Rebecca De Mornay) threatens Solomon (Ernie Hudson) in *The Hand That Rocks the Cradle*

anxiety is stirred by Solomon's unexpected appearance; a whole series of uncomfortable racial assumptions are awakened momentarily by his arrival. Yet once acknowledged and understood, and incorporated appropriately within the workings of the white house, Solomon becomes the appreciative recipient of benevolent white liberalism. However, Solomon's incorporation is always a partial one, susceptible to Peyton's manipulation of the Bartells' residual "common sense".

What is also striking about the movie is the total absence of any notion of Solomon's origins. A lone black man, Solomon is delivered into the heart of white Seattle by the geographically abstract Better Day Society, eliding in the process any notion of a black community. When Peyton accuses Solomon of sexual transgression, he is simply loaded back into the Better Day Society's truck, and whisked away. Solomon is more a figure of parable than a living and breathing African-American. The complexity of resonance provided by his name – a mentally challenged black man with the moniker of the wise King of Israel; an antebellum-period name with which to arrive at the Bartells' big white house, reveals Solomon as a semantic marker, invoked to impel the white family to question their assumptions. The Seattle drawn upon by *The Hand that Rocks the Cradle*, and used to generate an effective and evocative thriller, is not "a multiracial society in which Latinos, African-Americans and Asians are key players", but rather "a calmer, more culturally homogeneous environment", where, armed with vigilance against the white extremism implied by Peyton's extreme whiteness, a racially liberal status quo may conceivably be maintained, and thus white anxiety suitably quieted.[49]

FRASIER: WHITE FOR COMEDIC EFFECT

Six months after the Rodney King 'not guilty' verdict, and as the news media continued to profile Seattle as a white-flight centre, Los Angeles' culture industry released another highly successful and also critically lauded representation of Seattle, in the form of the NBC prime-time sitcom *Frasier*.[50] Like *Twin Peaks* and *Northern Exposure*, *Frasier* was also an example of 'quality television', used by the networks as a means of attracting "an audience with blue chip demographics … upscale, well-educated, urban-dwelling, young viewers advertisers so desire to reach".[51] Indeed, one of the co-creators of *Frasier* was James Burrows, a former director of *The Mary Tyler Moore Show*, widely understood as the progenitor of 'Quality Television' in the 1970s.[52] Burrows was also one of the creators and producers of the award winning prime-time sitcom *Cheers* (1982–93), from which *Frasier* was a 'spin-off'. *Frasier* relocated one of *Cheers'* large cast of regular characters, namely the cerebral Dr. Frasier Crane (Kelsey Grammer) from the original *Cheers* location in Boston, to his 'hometown' of Seattle. As Peter Casey, another of *Frasier's* co-creators, openly admitted, no episode of *Cheers* had ever mentioned Seattle as Dr Frasier Crane's hometown; indeed it had been invented post hoc, and reflected an enterprising opportunism on the part of *Frasier's* creators to associate themselves with a fashionable urban locale.[53]

The comedy in *Frasier* is centred upon the trials and tribulations of psychiatrist Dr. Frasier Crane, the host of a Seattle radio call-in advice show. The first few episodes of the show established Frasier's domestic circumstances, as his stylish and expensive upscale Seattle apartment was turned upside down with the arrival of his blue-collar ex-cop father, Martin, who was forced to move in with Frasier after being injured in the line of duty. Soon to follow Martin into Frasier's apartment was his live-in home-care provider Daphne Moon. Completing the ensemble of main characters was Frasier's brother Niles, also a psychiatrist and similarly upscale and effete in ideas and taste, and his cynical radio show producer Roz Doyle. The frequent visits of Roz, and most especially Niles, to Frasier's apartment constituted one of the main sources of *Frasier*'s 'comedy of manners'-style entertainment, and frequently generated storylines shaped by the characters' neuroses, anxieties and competition over taste, etiquette, social status, friendships and love lives.

What I want to draw upon here in particular is the notion of *Frasier* as a space for the articulation of modalities of racial anxiety. The first thing to point out is the fact that *Frasier* contained no non-white characters did not make it unusual for an American sitcom, even if the sitcom is one television genre that has a significant history of representing people of colour. This is not to say that those representations have not been deeply problematic; from *Amos 'n' Andy* (1950) to *The Hughleys* (1998–), concern over stereotyping persists. However, it is also fair to say that sitcoms do not need to be set in Seattle in order to elide non-white Americans. This is even more the case for 'quality' sitcoms, aimed predominantly at white upscale urban-dwelling young viewers. Yet even in comparison with arguably its closest counterpart, namely the Manhattan-set sitcom *Friends*, *Frasier* was striking in its racial and ethnic homogeneity. Unlike *Friends*, *Frasier* had no characters of even signified Jewish or Italian ethnic descent. In contrast, the show obtained its regular displays of comedic 'otherness' from Daphne Moon (Jane Leeves), a live-in physiotherapist for Martin (John Mahoney), Frasier's father.[54] Daphne was signified as a Mancunian, and the show (particularly in the first season, when the comedic possibilities of Daphne's Englishness were more novel) made much use of the eccentricities and quirks provided by her 'otherness'. It is also worth noting that Daphne's Englishness provided a less politically incorrect object of comedic attention, eliciting none of the potential disapprobation for mocking people of colour or other white ethnicities.

Frasier's discomfort in dealing with the issue of white ethnicities was apparent in the few episodes that did attempt to generate humour from such identities. For example, in the episode 'Merry Christmas, Mrs Moskowitz' Frasier is set up on a date with a young woman by her mother, Helen

Figure 8: Seattle nervosa: the cast of *Frasier*, NBC's Emmy award-winning sitcom

Moskowitz (Carole Shelley).[55] Frasier first meets Helen in a downtown department store, whilst shopping for a menorah (a lamp lit during the Jewish festival of Chanukah). Frasier's purchase is for his son Freddie, who is half Jewish, and lives with his mother, Frasier's ex-wife Lilith, in Boston. Helen spots the menorah, and assumes that Frasier is Jewish. A stereotypical forthright Jewish mother, Helen quickly sets up a date for her daughter with Frasier – in her words, "a nice unattached Doctor".

The comedy of mistaken ethnic identity ensues when Helen's daughter Faye (Amy Brenneman) visits Frasier's apartment, sees the Crane family's Christmas decorations, and realises Frasier's 'real' non-Jewish identity. With Helen due to arrive at Frasier's apartment, Faye persuades Frasier, who in turn persuades his brother Niles (David Hyde Pierce), and father Martin, to "pretend to be Jewish" long enough to fool Helen, who is scheduled to fly home to Florida later that day. What such 'pretence' amounts to is Frasier, Niles and Martin 'hamming up' (albeit gently) every stereotypical Jewish trait, mannerism and phrase they know, and misplacing aspects of Judiasitic ceremonial practice. The point is that "to be Jewish" for these Seattle characters is to seek recourse to a set of clumsy, half-formed clichés concerning Jewish ethnic identity. In other words, to be Jewish is to be 'other', and by association, that which is not encountered everyday in Seattle. It is no accident that Helen is from another city, a fact that suggests that complicated notions of ethnicity reside elsewhere, and allows non-ethnic whiteness to be restored by the end of the episode.[56]

'Merry Christmas, Mrs Moskowitz' is also interesting in the sense that the humour it seeks to generate from ethnic mimicry places it within a well-established American comedic tradition. As Susan Gubar points out in *Racechanges: White Skin, Black Face in American Culture*, although cross racial mimicry dates back to nineteenth century minstrelsy, it has been central to twentieth century American culture.[57] Gubar notes anecdotally the "exuberant hilarity that [comes] in part from the transgressive pleasure" of racial parody for white liberal academics who otherwise respect the right of minorities to represent themselves.[58] Gubar's book, reflecting the debt it owes to Frantz Fanon's seminal *Black Skin, White Masks*, concerns itself exclusively with black/white racechanging, yet her description works equally well for *Frasier*'s Jewish mimicry, (acknowledging the differing histories of African-American and Jewish (and Black Jewish) experience and representation that gives such mimicry its cultural resonance). Indeed, considering the fact that Frasier and his brother Niles are psychiatrists, there is an irony in the show's depiction of them in a process ethnic mimicry that, as Gubar points out, serves to "illuminate the psychology of whites".[59]

Indeed, not to disappoint academics interested in black/white racial parody, *Frasier* obliged in an episode entitled 'Something about Dr. Mary', the only one out of the first seven seasons to include an African-American in a leading role.[60] The episode begins with Frasier and Roz (Peri Gilpin), the producer of his local radio show (where he dispenses psychiatric advice to phone callers) having a conversation in their local coffee bar. Roz is going to be absent from the show for a short period, whilst having her apartment redecorated, the proof of which is the number of paint colour 'tester' cards she has arranged on the table. She looks at the cards and states wearily to Frasier that "after a while, you can't even tell the colours apart". Frasier looks at what appear to be three identical white cards, and proclaims smugly, "Eggshell, Ecru, and this, of course, is Navaho white." Roz looks at him with a wry grin, and suggests that he "tr[ies] turning the cards over" to the correct, painted side. A minor gag perhaps, but also a subtle preamble that anticipates the episode's concern with admitting the presence of "colour" differences.

As the scene in the coffee shop continues, Frasier tells Roz that he is "thinking of reaching out to the community". Frasier states that he was "guest speaker at a program called 'Second Start' for career training for people who are stuck in tedious, low paying jobs", and he is going to invite one of the trainees to produce his radio show in Roz's absence. If 'Second Start' sounds like a close relation of *The Hand that Rocks the Cradle*'s Better Days Society, then the ensuing scene, in which Frasier shows Mary Thomas (Kim Coles), a young African-American woman, around his radio studio serves to underline the similarity. Like the Bartells, Frasier is engaged in an act of benevolent white liberalism. As he compares his time at Harvard with Mary's night school experience, Mary notes sardonically that "we are practically separated at birth" (sic). Things get more interesting as the scene continues, with Mary taking her first attempt in the producer's booth. Mary is reluctant to speak 'on air' and Frasier, comfortable and relaxed with the rather paternal 'master and pupil' power relations, and the seemingly meek and silent black woman, encourages Mary to speak up. Speak up she does, transforming into a witty, 'sassy', brash and opinionated foil to Frasier's professional and authoritative persona. Interrupting the doctor's psychiatric advice to dispense her own 'home-spun' 'common sense' knowledge, Mary moves Frasier's self-consciously cerebral show closer to the tone of a confrontational television chat show. Mary's undermining of the hierarchy Frasier has put carefully in place is complete when she asks a caller to refer to her as Dr Mary, thereby dissolving the difference between the pair's status.

Returning to his apartment, Frasier vents his frustration at what he feels is Mary's usurping of his show. When questioned by his brother Niles as to why he has not approached Mary with his concerns, his father quips, "because she's black". Frasier denies vehemently the suggestion, stating that "it's a difficult situation," in the process indicating his liberal guilt over appearing, or perhaps unconsciously being, racially prejudiced. In exasperation, Frasier requests that Niles join him in a role-play of a hypothetical conversation between himself and Mary. Suggesting to Niles that "why don't you play me, and I'll play Mary", Frasier then imitates Mary, complete with signified mannerisms, deportment, intonation and 'sassy' attitude, stating that:

So you want me to stay in my place, massah! What, am I getting too uppity for you? … You have no idea how difficult it is for a black woman in a white man's world.

From the audible increase in studio audience laughter that accompanies Frasier's racial parody, it is clear that the "exuberant hilarity" and "transgressive pleasure" of the Waspish Doctor mimicking an African-American woman is being promoted. Yet as the cerebral Frasier mimics the sassy and outspoken Mary, the show locates itself within "a long tradition of identifying the European with the mind [and] the African with the mouth".[61] Moreover, Niles' insistence that Frasier's 'Mary' is saying things that the real Mary "would not say" suggests that Frasier's parody has moved away from mimicking an individual, to generating an archetypal black female. Susan Gubar asks rhetorically in *Racechanges* "how can white people understand or sympathise with African-Americans without distorting or usurping their perspective",[62] before admitting that the notion of "their perspective" is a problematic generalisation. Frasier's performative blackness sees him slip seamlessly into "their perspective", and in the process usurping black subjectivity.

Frasier's racial mimicry is made more complex by the fact that it is also a cross-gender parody, particularly since his character's persona combines brazen heterosexual amorousness with a stereotypical effeminate gay male subjectivity for 'tasteful' furnishings, fabrics, and so forth. As Gubar points out, some "racial masquerades … abrogate definitional limits. In various contexts and media, racial representations teeter between identification … desire … and disavowal".[63] Frasier arguably combines all three positions in his continual masquerade of homosexuality, before even beginning to consider his racial masquerade. Yet ultimately, despite the subversive potential of Frasier's 'racechange', the status quo is restored. By the end of 'Something about Dr. Mary', Frasier admits his misgivings to Mary, who in response tells Frasier that her appearances on his show have been so popular that she has been offered her own show on another station, adding "God bless your guilty white ass." Yet Mary's 'success' is also the episode's route to resolution, as blackness is displaced again from *Frasier*'s Seattle.

If *Frasier*'s ethnic mimicry amounts finally to nothing more than the desire for what Stuart Hall calls "a taste of the exotic … a bit of the other",[64] then as Gubar points out, it does at least make "whiteness as startlingly visible as blackness has historically been".[65] Indeed, *Frasier*'s Seattle never appeared more waspishly white than in these two episodes that attempted to deal with ethnic otherness. Gubar's suggestion that cross-racial parody is often linked with "white insecurity as well as status anxiety", is played out in Frasier and *Frasier*'s awkward handling of "race matters". [66] The sudden appearance of a black woman "in a white man's world" impels Frasier to fear not only his own racism, but also the threat to his status. That *Frasier*'s resolution allows both anxieties to be assuaged is made all the more apposite by its signified setting in a key white-flight city of the 1990s.

Like *The Hand that Rocks the Cradle*, *Frasier* was a high-profile, highly successful representation of Seattle that was also particularly responsive to the wider mediation of the city's delimited racial demographic. Both *The Hand that Rocks the Cradle* and *Frasier* emerged at precisely the moment when Seattle's identity as a key white centre was being mobilised within the mainstream news media. It is important to recognise that this did not compel such narrativised representations to simplistically 'reflect' this wider discursive context, but rather that they chose to draw upon the timely meanings of Seattle's whiteness and shape them within their own generic representational parameters. As a thriller, *The Hand that Rocks the Cradle* used notions of Seattle's whiteness to construct a plausible landscape in which to generate a suspenseful narrative that revealed 'white anxiety' at the heart of outwardly benevolent white liberalism, and thus worked to contest and challenge its self-satisfied veneer. Whilst *Frasier* exemplified the general over-representation of Seattle as a 'white city' within the mainstream American media, in the episodes examined here it could be seen on occasion to utilise middle-class 'white anxiety' within a comedic mode of suspense. As Seattle's role as a key white flight centre suggests, the city had an important role to play within wider discussions and debates about race and urban centres, particularly in the early 1990s. As key texts in the signification of Seattle in the 1990s, and important popular representations in their own right, *The Hand that Rocks the Cradle* and *Frasier* thus also provide a means of expanding what has been a fairly restricted repertoire of generic forms and represented urban spaces used to explore the role of America's cities as "more intensely psychologised as sites for racial anxieties" in the 1990s, and which have up until now remained primarily focused on Los Angeles and New York.[67]

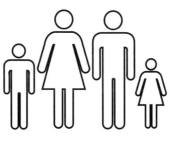

CHAPTER 4

SEATTLE: RATED SUITABLE
FOR ALL THE FAMILY

Why should I not love Seattle! It gave me a home by the beautiful sea, it brought abundance and a new life to me and my children and I love it, I love it![1]

It was a morning when, shootings aside, Seattle really is one of the most livable cities on the face of the earth.[2]

The late 1980s witnessed an upsurge of articles in the American press devoted to cities understood to be undergoing 'revitalisation' – a byword for the emergence of the converted warehouses and wharves that signified the presence of gentrifying 'urban pioneers'.[3] Yet as Robert Beauregard notes, "middle-class commentators wanted more, however; they wanted liveable cities, urban counterparts to gentrified neighborhoods".[4]

One urban centre that became a near-permanent feature of such commentator's lists of liveable cities was Seattle. Yet what was particularly interesting about the profiles of Seattle was the fact that they tended to emphasise the fact that it was a desirable place to bring up the middle-class family. For if liveability profiling was in large part about signalling that particular cities were attractive to the middle classes, it is important to note that it tended to project an even more circumscribed vision of 'urban living'. In particular, profiles were inclined to be skewed towards single individual households and childless couples in the professional-managerial sector, reflecting the fact that, as Neil Smith and Peter Williams point out, the transformation of city cores into sites for corporate headquarters and business services coincided with shifts in the labour force.[5] In particular, the continuing upward trend in the number of middle-class women choosing to establish careers meant the postponement of families, and the increase of childless individuals and couples with more disposable income, to whom the new consumption-oriented lifestyle profiles of 'city living' were most obviously aimed.[6]

In contrast, Seattle's profiling as a desirable city for middle-class family life gave it a particularly compelling appeal. Crucial to the post-war history of the American city was the movement of middle-class households out of the cities and into the suburbs, in the process depriving the city of "taxpayers, workers and consumers".[7] This exodus of middle-class families from the city was pivotal to mainstream narratives of 'urban decay and decline' right into the 1980s, whilst the profiling of 'edge city' development in ex-urban areas across the United States referred to in chapter one provided another key element of the discourse on urban decline.[8] In this way, the talismanic value of the 'middle-class family' as a symbolic touch-stone for the health of America's cities was quite clear. Indeed, articles that tentatively pronounced the reversal of urban decline would cite the trickle back of middle-class families into particular cities as the ultimate and irrefutable proof of liveability bucking the inexorable trend towards crisis and decay.[9]

In this chapter I examine selected texts from the period that were especially responsive to Seattle's identity as a city for middle-class families. As we shall see, images and narratives of the city's 'liveability' did not confine themselves to the national print media but also found themselves worked into other types of representation, in particular some of the most popular and critically acclaimed cinematic and televisual portrayals of the city. These representations drew upon notions of Seattle's liveability in complex ways, transforming the concept as they worked it into fictionalised forms.

ORIGINS OF THE LIVEABLE CITY

As the discussion of the historic conflict between countryside and city in chapter two points out, the notion of questioning whether cities are fit to live in and thus in the broadest sense 'liveable' is nothing new. Moreover, the preceding chapters have reflected upon the different ways in which a number of Seattle's key representations connected the city to enduring themes pertaining to urban vitality and well-being. Implicit in many of those representations were a range of urban qualities and characteristics that also coalesced around the concept of the 'liveable city'. Whilst the profiling of 'liveable cities' was particularly prevalent from the late 1980s onwards, it is important to note that the American print media had been concerned with the subject from the mid-1970s onwards, and that Seattle had been high amongst the contenders from the start. Magazines such as *Time, Harper's, Atlantic Monthly* and the *Christian Science Monitor* all published articles that sought to assess the 'liveability' of various American cities. Two factors in particular can be said to be crucial to the way in which cities were discussed, and understood subsequently to be or not to be liveable. The first was the wider discursive framework within which more general debates concerning the health of American cities were being situated at that time. The second was the relatively narrow band of class-based and political affiliations represented by commentators in the national print media. Together, these two factors served to create definitions to determine and portraits to represent the so-called liveable cities.

From the outset, the rhetorical logic of the search for liveable cities suggested that the bulk of cities were assumed not to be so, in effect pointing to a subsumed identity within the broader discourse on urban decline. Indeed, the very first of such articles, published in the January 1975 edition of *Harper's* magazine was entitled 'The Worst American City'. Its author, Arthur M. Louis, located his article's paradigmatic origins in a study by H. L. Mencken for the *American Mercury*, published in 1931. Mencken had sought to identify the 'worst' American state through the process of computing each state's average ranking in more than one hundred statistical tables.[10] Louis justified the similarly negative nature of his title by proclaiming that "it doesn't require any perversity of character to go looking for the worst city. There are no good cities in America today – only bad and less bad".[11]

It should be remembered that Mencken was conducting his study of the American states in the midst of the Great Depression, which provided a reasoned rationale for his pessimistic philosophy. Yet Louis' study of American cities was completed against the backdrop of a somewhat comparable scenario – the nation was just beginning to emerge from the worst economic recession since before the Second World War.[12] Significantly, this was a state of affairs understood to have brought a number of the nation's large cities "to the brink of fiscal crisis", as the industrial output on which they were dependant went into steep decline.[13] Moreover, as recently as 1972 the Commission on the Cities, a key governmental body set up "to assess changes since the Kerner Commission Report on the riots of the 1960s" had lambasted the "shameful conditions of life in the cities … housing is still a national scandal; schools tedious and turbulent; crime, unemployment, disease, heroin addiction higher, welfare rolls longer".[14]

Against such a backdrop Louis' search for the least worst city in America seemed decidedly optimistic. Nevertheless, using 24 statistical tables, including figures for crime incidence, health care provision, housing quality, high school graduation percentage, and the number of public amenities, Louis came to the conclusion that the least worst city in America was Seattle. Seattle achieved this, in the author's eyes, not necessarily by being the best in every category, but rather by being always amongst the least worst. Having seen where his statistics had landed him, Louis concluded, in the process admitting his own disposition: "Seattle's position at the top of the list seems plausible enough. As one who has been there, I can attest that it is an appealing and immensely civilised place."[15]

Had Louis attempted to conduct his study four years previously, it is unlikely that he could have reached the same upbeat conclusion. Seattle had by no means been exempt from the ravages of the national economic slump. Between 1968 and 1971 the city suffered tremendously from the recession at the aeroplane manufacturer Boeing, the metropolitan area's largest and most influential employer. A well-documented range of factors enabled Seattle to make the transition from such a state of affairs. As the recession abated, Boeing made a series of steps towards diversification and into electronics and industrial research. The Port of Seattle experienced its most successful period to date turning its investment in expansion and modernisation.[16] Going hand in hand with such activities was the physical development of the city's downtown, as apartment houses, business offices and bank-financed skyscrapers were constructed.[17]

Also pivotal to the revitalisation of Seattle's downtown was the increased emphasis on new or regenerated sites for consumption. Department stores, shopping centres, and a new major league sports stadium were all constructed. Two of the city's oldest sections, Pike Place Market and Pioneer Square, were given over to revitalisation and were designated historical districts, creating in

Seattle for the first time what the historian Roger Sale called an "honest-to-God tourist trap".[18] Revealing his own opinion of this new landscape of leisure and consumption, Sale offered that Seattle "has become a great place in which to buy things".[19] After the recession, articles in the nation's news periodicals seemed to indicate that the centres of a number of the nation's large cities were being "rediscovered" by the middle classes.[20] Articles in the national print press profiling such liveable cities worked to mediate the processes of urban revival for middle-class Americans, creating positive narratives and images out of the reconstructed urban topography, and Seattle found itself at the forefront of this range of cities being lauded for their attractive environments.

It was no coincidence that a proliferation of articles attempting to *rank* American cities occurred at precisely the time when a number of them were re-structuring, in the process reflecting the increased importance of attracting multinational corporate headquarters and also a more prosaic repertoire of business services. The notion of ranking cities in descending order paid homage to that competitive ethos, and to the kudos of 'being number one'. It is also appropriate that in a period in which the nation's cities were adjusting to the new regime of 'flexible accumulation', characterised by what David Harvey terms "new ways of providing financial services, new markets ... greatly intensified rates of commercial, technological and organisational innovation", and multinational capital flows, that the method by which a city was often to be judged was on its numeric performance.[21] In Louis' study, Seattle was the least worst because its average score was 14.0, whereas Newark, for example, rated a lowly 41.6. To reduce a city's social, political and economic complexity to a digital abstraction not only served to elide the significant differences between cities, but also the widening gap between a city's richest and poorest members.[22] For example, it is apposite that *Time* magazine's own liveable cities chart for 1975 entitled, imaginatively, "Ranking the Cities" was published next to the list of the New York Stock Exchange closing prices. In this way, it took only a sideways glance to note that if General Electric had gone up a point, Seattle had gone up too.[23]

If the proliferation of numeric city ranking charts could be understood as correlative to the stocks and bonds figures which shaped increasingly the cities' economic future (in Seattle the Bank of California and the Seattle First National Bank both constructed imposing skyscrapers, thus competing vertically for symbolic pre-eminence), then they were not at the expense of more conventionally narrated accounts of liveability. If one article could be said to encapsulate the various elements of these liveability profiles, it was "America's Most Livable City: Seattle", written by Judith Frutig for the 21 May 1975 edition of the *Christian Science Monitor*.[24] Frutig's article was a lengthy 'lifestyle'-section piece on the city, and used short profiles of Seattle inhabitants as the main technique with which to construct a portrait of the city. Firstly, it is worth noting how obviously skewed the author's portrait of the city was towards foregrounding its professional, middle-class inhabitants. For example, the first short portrait she includes is of Ralph Perry, who commutes by ferry from his home on Bainbridge Island to work in downtown Seattle. Frutig enthused over Mr Perry's commute: "he suffers neither fumes nor frustrations. He is never hung up in traffic – except when the whales run" – an idyllic journey that the author notes is available to "a large portion of [Seattle's] professional population". Indeed, Frutig noted that commuters complained when the ferry journey time was reduced, since it meant "they had no time to finish the newspaper".[25] It would be hard to conceive of a way in which Frutig could have commenced her article with a more middle-class portrait of the city than with this paean to the joy of commuting professionals.

From here Frutig did not turn to consider what life may be like for Seattle's refuse collectors or restaurant workers, but went straight to Edward Sullivan, a member of the Seattle Chamber of Commerce and Jere Bacharach, an associate professor at the University of Washington. Like Mr Sullivan, Mr Bacharach was identified as a family man, and Frutig described the Bacharach family's weekend journey from their "ranch house, overlooking Lake Washington, where they can fish for salmon off the dock" to Pike Place market, to purchase "two Dungeness crabs, a salmon, a pound of shrimp, and some fresh vegetables for a homemade seafood dinner". To consolidate the ethos of tasteful consumption, Mr Bacharach noted that the pleasure of Seattle was the "good cultural feeling. To me, culture is everything."[26]

The article's concerns and biases towards middle-class lifestyle and con-sumption are, of course, no surprise, reflecting as they did the imagined concerns and aspirations of the magazine's readership demographic. It is also evident that the magazine anticipated the more extensive profiling and dissemination of Pacific Northwest 'lifestyle' that would emerge in the 1990s.

Also significant is a couple of the piece's more subtle and inferred meanings. The first is the fact that two of the inhabitants the author focuses upon – Mr Sullivan and Mr Bacharach – were men with families who have re-located to Seattle. Mr Bacharach, from Long Island, noted that "we've gotten so that we don't like to go back East anymore. After a couple of days of that tension, I wonder how folks can stand it 24 hours a day."[27] As such, the Bacharach family was able to attest to Seattle's comparative virtue, and more broadly, to the desirability of geographic mobility. Indeed, the article was distinguished by descriptions and photographs of ferries carrying businessmen; the Monorail whizzing by; car journeys downtown; all motifs of pleasurable movement that helped reinforce the ethos of middle-class mobility. In doing so, the article was in keeping with the ideological underpinnings of liveability rankings – namely that cities could be adopted or left behind as their rating points wax and wane, and served to valorise a way of life that foregrounded 'lifestyle' choices.[28]

Apparent from Frutig's article is the fact that liveability profiling was from the outset the overwhelming preserve of national periodicals that sought to profile urban vitality for a middle-class readership demographic. Liveability, in this way, was most often about appealing to a geographically mobile white professional middle-class that was perceived to be able to choose where to live, largely on the basis of selected lifestyle criteria.[29] Nowhere was the circumscribed nature of the liveability profiles made clearer than in a *Time* magazine article from December 1977. Stating that the city was "bustling, clean and eminently livable", *Time* noted also that Seattle had "too few blacks for any real racial problems".[30] Although *Time* was quite exceptional in its decision to make explicit the fact that urban decline functioned frequently as a euphemism for race, what was not at issue, as the last chapter points out, was that urban vitality and liveability would continue to function as signifiers of Seattle's racial homogeneity into mainstream media accounts of the 1990s.

VOICES OF DISSENT: CRACKS IN SEATTLE'S LIVEABLE IMAGE

Throughout the 1980s, articles dedicated to profiling America's most liveable cities proliferated. This is perhaps no surprise, since the key buzzwords in the 1980s for commentators concerned with the state of the cities were "revival, revitalisation, renaissance and rediscovery".[31] A host of articles waxed lyrical on the increased investment in central business districts and the expansion of corporate and retail services.[32] Yet commenting on this phenomenon and the promotional literature that accompanied it, David Harvey notes the "grim history of deindustrialisation and restructuring that left most major cities [with] few options except to compete with each other, mainly as financial, consumption and entertainment centres".[33] The emergence of cities characterised by widening extremes of wealth and poverty, manifesting social division and patterns of uneven development did not go unnoticed; as Beauregard states, many commentators "found it difficult to overlook an increasingly visible, bothersome, and deepening bifurcation of the city into rich and poor".[34]

In this section I want to pay close attention to three representations that fixed upon Seattle as an example of just such a bifurcated city. Not only are these representations important instances in the signification of Seattle – they all received high critical acclaim – but they also comprised a sequence of interconnected counter-narratives to Seattle's prevailing liveable image. As such, they signified noteworthy moments of contestation with the 'liveable' profiling. The first representation is a photo essay from *Life* magazine in July 1983, entitled 'Streets of the Lost', written by Cheryl McCall, and with photographs by the award-winning photographer Mary Ellen Mark.[35] 'Streets of the Lost' provided the impetus for the second representation to be considered, the Academy Award-nominated documentary *Streetwise* (1984), produced by Cheryl McCall and directed by Martin Bell. *Streetwise* was in turn the inspiration for the third representation under consideration, namely the critically acclaimed feature film *American Heart* (1992), co-written by Cheryl McCall, produced by Mary Ellen Mark and again directed by Martin Bell. Whilst I explore the ways in which the representations provided moments of contradiction that undermined Seattle's liveable image, I also want to use the contrasting fortunes of *American Heart* and its cinematic counterpart *Sleepless in Seattle* (1993) to illustrate the ultimate hegemony of mainstream liveability.

Life was a long-standing mass-circulation general interest magazine, most renowned for its photojournalism. Part of the *Time* magazine group, *Life* ceased regular publication in 2000, largely a victim of the market trend towards more specialist magazines. However, at the time of the publication of 'Streets of the Lost' in 1983, *Life* was still a significant monthly publication, and, with *Time* and *Newsweek,* one of a cohort of mass-circulated general-interest periodicals. *Life*'s editorial policy placed an emphasis upon articles and images that 'recorded' key moments in the life of the nation, a policy that was shaped by a broadly 'ideologically centrist' notion of events of national interest. As well as recording public events, *Life* also carried photo-essays of a more intimate scale, which were nevertheless framed by an editorial that sought to connect them to narratives of national relevance and import. It was with the latter category in mind that *Life* published 'Streets of the Lost'.

'Streets of the Lost' was a photo-essay depicting the lives of young homeless kids on the streets of Seattle. The text was written by the journalist Cheryl McCall, who was a regular contributor to *Life*. The photographs were by Mary Ellen Mark, widely recognised as one of America's most important documentary photographers and noted for her portraits of harsh urban environments.[36] Against the sub-heading 'runaway kids eke out a mean life in Seattle' the article used a series of Mark's stark black and white photographs depicting homeless kids holding guns, being arrested, injecting drugs, rummaging for food in trash bins and conversing with pimps.[37] What is evident from McCall's accompanying text is that the article was not seeking to suggest that Seattle was unusual in possessing 'Streets of the Lost'. On the contrary, what was being stressed was Seattle's analogous identity with other American cities. Seattle was depicted as a typical urban landscape scarred by prostitution, drug dealing, and physical and sexual violence, made clear in the essay's assertion that the plight of homeless kids was a problem that affected "every city in America".[38] However, it is arguable that when considered alongside its sister publication *Time*'s propensity to reward the city for its liveable qualities, the decision to select Seattle rather than New York or Chicago would have served to generate a more shocking piece of *verité*; in other words, that the city being lauded concurrently in other national periodicals was here to be stripped bare and exposed behind a liveable facade.

What made the *Life* article more interesting was that it focused on the plight of children. This obviously had a useful function in terms of generating a powerful 'human interest' angle to the article – focusing on the vulnerability of children framed urban decay in a most evocative manner. However, it is perhaps more accurate to say that the *Life* article predominantly used the city as an emblematic site – what Beauregard terms the 'spatial fix' for situating more widespread anxieties – in this case the breakdown of the American nuclear family. The article was most explicit in seeking to foreground the breakdown of the family as the key instigator of the problems to be found on the streets of Seattle, noting that "kids ordinarily don't run away from warm, loving families".[39] 'Streets of the Lost' depicted kids "fleeing turbulent households racked by conflict, violence, neglect and – in a disturbingly high percentage of cases – sexual abuse". Moreover, it was careful to highlight the diverse family backgrounds from which these street kids originated. Erin's unemployed mother and stepfather live "in a one-room apartment over a tavern in downtown Seattle", whilst Sam, 17, is "a professor's daughter from Idaho".[40] The *Life* article thus drew Seattle into a powerful strand of social commentary that was gathering increasing momentum as the 1980s progressed, namely, the debate over the state of the American family.

Not unlike the nation's cities, the state of America's families has often been regarded as a symbolic touchstone for appraising more broadly the health of the nation. Moreover, during the 1980s and 1990s both garnered a range of popular and academic commentaries willing to diagnose both as endemically sick and in need of serious repair. Indeed, it is worth noting that the city and the family have on occasion been implicated together in very high profile national narratives of decline.[41] On any list of the most infamous of these discursive intertwinings would be George Bush's 1992 State of the Union address, in which the President claimed that "the crisis of the cities results from 'the dissolution of the family'".[42] In line with a host of previous conservative commentators, Bush made the "breakdown" of family stability in the city the root cause of a range of social ills, such as poverty, crime, homelessness and drug abuse. Bush's words reflected the fact that, as Arlene Skolnick notes "since the early 1970s, public conversation about family change has been dominated by a bleak rhetoric of loss and decline".[43] Skolnick points out that in spite of a number of scholars of the family arguing "that the decline of the family is a myth"...[44]

> ... by the late 1980s, the optimists themselves were becoming an endangered species. The issue was no longer whether the family had a future; the break-

down of the family seemed firmly at the centre of a broad syndrome of cultural decline.[45]

The 'breakdown' of the family was therefore seen as both indicative of and instrumental in more broad processes of 'decline'. Such a 'commonsense' understanding found itself worked into representational form in 'Streets of the Lost', which sought to mobilise Seattle as a 'typical' American city, and therefore an obvious site of social strain and fracture. Thus whilst 'Streets of the Lost' had much in common with the body of writing and reportage that flourished in the 1980s and 1990s known as 'underclass ethnography' – most particularly its formal similarity in creating the impression that the reader/viewer was experiencing the eyewitness account of an 'objective' and 'detached' observer – it was also thematically distinct.[46] This was most apparent in the way that it emphasised the connections between the street children and (albeit terminated) suburban middle-class family origins, rather than suggesting, like much of the contemporaneous reporting of the 'underclass', that those being viewed were part of a separate and distinct social grouping.

A similar combination of formal and thematic influences could be seen at work in *Streetwise* (1985) a documentary film produced by Cheryl McCall, and directed by Mary Ellen Mark's husband, Martin Bell, formerly a leading cinematographer of nature programming.[47] Inspired by 'Streets of the Lost', *Streetwise* was nominated for best documentary feature at the 1985 Academy Awards, and drew much critical acclaim for its ostensibly honest, candid and unmanipulative cataloguing of the lives of the same group of street kids photographed for the *Life* photo essay. *Streetwise* blended unobtrusive hand-held *verité*-style camerawork with long-shot footage that owed much to Bell's experience filming wildlife documentaries. Given the pejorative history of the use of the 'urban wilderness/urban jungle' metaphor within popular narratives of the American inner city, it could be wondered whether the use of such techniques revealed an implicit or perhaps even unconscious analogy. However, a reasonable assessment of *Streetwise* would suggest it was a sympathetic portrayal of street kids, even if the seemingly artless style and plotless interweaving of footage of the kids' daily lives was not quite as it seemed. Commenting on *Hoop Dreams* (1994), a highly acclaimed documentary film of black urban life, Liam Kennedy points out that 'liberal investigation' documentary film-making, of which *Streetwise* was also an example, "create[s] the illusion of direct, unmediated engagement with the world of others".[48] This was certainly the intention of *Streetwise*, yet as the film progressed there was a palpable transition. The earlier sections of the film foregrounded the rootlessness of the kids and thus forced the viewer to connect individual behaviour to more abstract social forces and environmental influences; this was reflected in the extensive use of long-shot photography and distancing camera techniques. The latter sections displayed a movement towards the privileging of more conventional biographical micro-narratives that explained behaviour in terms of individual circumstances and unfortunate or often tragic events.

Nowhere was this more in evidence than in the film's depiction of a young boy named DeWayne. The film used the device of letting DeWayne recount his own story, as he outlined how his mother had abandoned him for a sailor, whilst his father had been put in prison. As the film moved towards its dénouement, it tightened its focus upon the various threads of DeWayne's narrative, including a couple of scenes in which the boy's obvious 'playing-up' for the camera marked a noticeable shift towards a more melodramatic mode of cinematic representation. Indeed, the film ended on arguably its most conventionally affective and powerful scenes; the first involved a conversation between DeWayne and his incarcerated father, as the father tried to implore his son not to follow his lead into petty criminality; the second showed the father, chained to prison guards, attending his son's funeral, after DeWayne had committed suicide in a Seattle detention-centre cell.

Perhaps the most compelling evidence for the way in which DeWayne's story moved *Streetwise* towards the potential for a more melodramatic and explicitly narrativised mode of cinematic representation was the fact that it provided the inspiration for the feature film *American Heart* (1992) co-written by Mary Ellen Mark, and directed again by Martin Bell.[49] In the latter stages of *Streetwise*, the film depicted DeWayne talking of his dream of going to Alaska with his father, and this was used to propel the narrative of *American Heart*. The feature film abandoned *Streetwise*'s group-focused, *vérité* approach in favour of a conventional, central-protagonist oriented, causally driven narrative focused around Jack and his son Nick – the fictional surrogates for DeWayne and his father. Generically, the film was thus constructed as a decidedly downbeat family drama – the first establishing shots of the film rendered the city a cold, grey space, inhabited by down and outs, buskers and prostitutes. Indeed, *American Heart* included many of the homeless teenagers who featured in *Streetwise*, yet this time merely forming part of the backdrop of flophouses, strip joints, bars and cheap hotels against which *American Heart* constructed its dramatic landscape.

The main premise of *American Heart* is that Jack, a petty criminal released from prison on conditional parole, comes back to Seattle seeking work, and to rebuild his life. He is followed by his young son Nick, determined to remain attached to his father, despite Jack's repeated entreaty for Nick to "take the bus back to Mount Vernon" and the Aunt and Uncle who look after him. Jack warns Nick that "I wouldn't be doing you no favours if I kept you with me" and that "I'd do a lousy job raising you". The movie makes clear that Nick is very much the casualty of a broken home, in the manner of the kids identified by the 'Streets of the Lost' article, and in keeping with the discourse of family crisis existent more broadly. In addition to having a convicted felon for a father, Nick has had no contact with his mother since birth. Nick carries with him a photograph of a woman he understands to be his mother, but finds out subsequently is actually a catalogue model, who bears some faint resemblance to the real person. To compound the blow, in the midst of a fight between father and son, Jack tells Nick that his mother was not in fact a model, as he had been led to believe, but rather "a five-dollar street whore". The implication is that Nick's idealised, cherished image of the maternal love he seeks is a deceptive simulation, and that right from conception, the boy's life has been the antithesis of the family life whose fragments he had hoped to re-unite.

The film's events are built around the central story of Jack's reluctant acceptance of Nick into his life, and his stuttering attempts to provide what he perceives of as a supportive paternal role. Jack allows Nick to stay with him at a flophouse room he has rented, whilst setting himself up as a window cleaner at a downtown skyscraper. Inhabiting the flophouse are similarly troubled and 'broken' families struggling to survive in the city. Of particular note is Molly, a teenage girl befriended by Nick, whose mother is a stripper at a peep show. In one scene the camera lingers while Molly and her mother argue – the mother telling Molly that she is a "fucking little selfish tramp"; the daughter retorting that "you're just jealous 'cos you're old and you're ugly and you're used up". Jack and Nick observe the scene, with the implication being that this is an inappropriate model for parent-child relations they should seek to avoid.

Jack's initially reluctant guardianship of Nick is in part due to his more pressing concern with finding the woman who had written to him in prison. Attempting to consummate this possible relationship as quickly as possible, Jack brings the woman, Charlotte (Lucinda Jenny) back from the seedy bar in which he had found her, to his flophouse room. Finding Nick there, he forcibly ejects him from the room. Charlotte balks at this display of cynical behaviour, but nevertheless finds Jack charming enough a few scenes later to take him back to her apartment. After this, some of the film's most touching scenes involve Jack's attempts to introduce Charlotte clumsily as a potential surrogate mother to Nick. Yet this is abandoned without resolution as the more pressing concern with obtaining money for the journey to Alaska takes precedent.

The story has an unhappy dénouement, in keeping with the ethos of tragedy that provided the initial impetus for the movie. Jack gets fired from

Figure 9: Nick (Edward Furlong) alone on the bleak streets of Seattle in *American Heart*

his window-cleaning job, as Nick gets drawn into the life of petty crime and truancy inhabited by the street kids of Seattle. Rainey (Don Harvey), Jack's former criminal partner, attempts to enlist the services of Jack, and then Nick, to help with his latest criminal ventures. The film ends with Jack being shot dead by Rainey, and grieved over by Nick, upon a ferryboat, thus echoing heartbreakingly the journey to Alaska they had dreamed of undertaking. The Seattle of the movie is a harsh, unforgiving environment, where broken and fractured families struggle to survive, kids fall into crime and vice, and lives are brief and brutal.

Yet *American Heart* contains one striking sequence that serves to rupture the harsh, colour-starved *mise-en-scène* that characterises the film, and underscores its bleak narrative. Towards the end of the movie Rainey persuades the now homeless Nick to help him undertake a burglary. Rainey drives Nick and a friend he has met on the streets to an affluent neighbourhood of Seattle. Nick smashes a door to gain entry to a well-kept detached house and in doing so figuratively breaks into another world. The scenes within the house are shot with the use of a different palette of colours – in particular warm, rich tones – and also with an extensive use of high key lighting that serves to create a marked contrast with the rest of the film. The effect is quite startling, as the release from the visual bleakness that characterises the rest of the film is combined with one of its bleakest sequences of storyline. Whilst in the process of burglary, the inhabitants of the house return, and we see that it is a father and his young son. The camera frames their movements as they enter the house, unaware of the intruders inside, and moves with them as they engage in affectionate father and son 'small talk' – the verbal signifiers of a balanced, role-appropriate filial relationship. As they stumble upon the intruders, the father moves to tackles Nick's friend. As the struggle ensues a handgun goes off and the son of the house is killed. Nick flees from the scene as the film reverts to its previous washed-out tones. In making the victims of the crime a father and son, the film is quite obviously setting up the scene for symbolic effect. By making such a chain of consequence between the impact of deleterious social conditions on the city's poorest families and their more affluent counterparts, the connection between families of different classes that characterised the original 'Streets of the Lost' photo-essay survived the migration across different forms of representation and into *American Heart.*

PHENOMENALLY LIVEABLE:
SLEEPLESS IN SEATTLE

Whilst *American Heart* received an overwhelmingly favourable critical reception, it was not a success with cinema audiences, and grossed a mere $325,220 in domestic box-office receipts.[50] That figure was put into perspective by the success of another motion picture set in Seattle released only a few months later. This second movie also concerned itself with the story of a father and his son involved in the search for the female partner/new mother who could 'complete' the family unit, and was equally explicit in seeking to foreground the theme of family life. *Sleepless in Seattle* (1993), a contemporary romantic comedy starring Tom Hanks and Meg Ryan went on to make $125,636,987 at the domestic box office, and was also a worldwide commercial success, whilst *American Heart*, despite an arguably more favourable initial critical reception, faded quickly from view.[51]

One of the most successful films of the 1990s, and by far the most successful film to be set in Seattle, *Sleepless in Seattle* was the story of Sam (Tom Hanks), a recently widowed father who moves from Chicago to Seattle with his son Jonah. One evening Jonah phones a national radio talk-show, with the intention of trying to find a 'new wife' for his dad, and thus in the process gaining a 'new mom' for himself. Sam gets drawn into the phone conversation, while Annie, (Meg Ryan) on her way home to Baltimore in her car, hears him speak, and starts to develop an obsession, first with the story, and then increasingly with Sam. This obsession, fuelled by watching the romantic movie classic *An Affair to Remember*, propels Annie to travel to Seattle in search of Sam and Jonah. Yet once she has located them, Annie loses her nerve and returns home to her fiancé Walter (Bill Pullman). Without yet knowing who she is, Jonah sends Annie a letter, stating that Sam will meet her at the top of the Empire State Building (mirroring deliberately the intended dénouement of *An Affair to Remember)*. After a series of comedic mishaps, the movie ends with Sam, Annie and Jonah meeting up, replete with romantic chemistry, on the observation deck of the Empire State Building.

A brief examination of *Sleepless in Seattle* illustrates precisely the way in which it constructed a quite different vision of father and son in the city to *American Heart*. Reflecting its generic identity as a bittersweet romantic

Figure 10: Perfect for family life: Sam (Tom Hanks) with his son Jonah (Ross Malinger) in *Sleepless in Seattle*

comedy, *Sleepless in Seattle* replaced *American Heart*'s terminally broken and damaged nuclear family with the image of the 'perfect' nuclear family damaged by tragedy. At the beginning of the film we are shown father and son in a graveyard outside of Chicago, attending the funeral of Maggie, the late wife and mother. The camera then pans up from the graveyard to reveal the Chicago skyline, skyscrapers looming large like monumental tombstones over the proceedings. The shot enables us to see that Sam and Jonah are not attending the funeral alone, but rather, are isolated from a group of mourners, on the opposite side of the grave. We are never given the specific details concerning Maggie's death; simply that it is a heartbreaking misfortune. In order to work as romantic comedy, it is important that the depiction of Maggie's death is not implied to result from anything that could be construed as insalubrious, such as AIDS, or shocking, such as murder. Thus it is also vital that to the movie's depiction of the nuclear family that Maggie *does* die, rather than divorce, abandon or cheat on Sam. In this way Maggie's 'sacrifice' preserves the family's innocent virtuousness.[52]

The way in which *Sleepless in Seattle* approaches the notion of finding a new mother/partner provides an interesting contrast with *American Heart*. The film creates much entertainment from depicting what it sees as the etiquette of modern dating rituals. Sam, as a formerly married man, has to learn what it means to date in the 1990s. Recalling the old gender roles of courtship, Sam asks his son "if it still works this way?" Jonah replies that "it doesn't – they ask you". Yet what such 'newness' actually prefigures is AIDS-era dating protocol, with an enforced cautiousness that impels delayed gratification, and "getting to know each other". The result is that new dating simulates the patterns of imagined old-fashioned 'romance' – as embodied in the movie's nostalgic evocation of *An Affair to Remember*. The effect is that Sam and Annie fail to share even the same screen space until the final scene of the movie, when they get to indulge in an affectionate linking of hands – accompanied by Jonah. In contrast, such protocols of modern dating do not seem to have reached the Seattle of *American Heart*, where after a few hiccups, Jack and Charlotte waste just a little time engaging in sleazy bar small talk before consummating their relationship.

Such differing approaches to courtship stratify along class lines. All the characters in *Sleepless in Seattle* are quite comfortably middle-class. In place of *American Heart*'s flophouse, Sam and Jonah live on an upscale houseboat in an expensive Lake Union marina. The houseboat is tasteful and comfortable, and a perfect family space – again, in contrast to *American Heart*'s flophouse room which is too cramped for Jack and Nick (who have to split up their single bed into separate base and mattress to uncomfortably accommodate them both). Sam is an architect and therefore a "skilled professional", unlike Jack in *American Heart*, who when asked by his parole officer to list his skills, replies resignedly that he "can't do nothing". It is also worth noting that Jack's return to Seattle is prompted by necessity – he is required to check in with the parole office situated in the city. In contrast, Sam and Jonah's move to Seattle is one of choice – almost of whimsy. In a pre-credit sequence in the movie, Sam is seen sitting in his Chicago office after the funeral, and expressing his desire for "a real change, a *new* city … I was thinking about Seattle". That Sam chooses Seattle as his 'new city' (a process befitting the commodification of cities by the rankings charts) underlines the city's 'discovery' in magazine lifestyle profiles and travelogues. Sam's class and economic standing make it viable for him to pick and choose the city in which he wishes to live, in a manner entirely according with the ethos of the 'liveability' profiles.

Clearly, one can attribute the pronounced disparity in the way that *American Heart* and *Sleepless in Seattle* approached the theme of family life in the city in large part to their generic differences. *American Heart* is a downbeat melodrama drawing much of its inspiration from *Streetwise*'s bleak urban ethnography. *Sleepless in Seattle* is one of a cycle of adult romantic comedies, including the Manhattan-set *When Harry Met Sally* (1989) written by Nora Ephron, and the Manhattan-set *You've Got Mail* (1998) written and directed by Ephron. *When Harry Met Sally* and *Sleepless in Seattle* were both 'sleepers' – what Thomas Schatz refers to as "mainstream A-class star vehicle[s] with sleeper-hit potential" – whereas *You've Got Mail* was an obvious attempt to repeat the success of *Sleepless in Seattle*, pairing again Tom Hanks and Meg Ryan.[53] That *American Heart* did not succeed in appealing to the American public could perhaps be attributed to the fact that it was so out of synchronicity with the prevailing mainstream accounts of Seattle. It was certainly not in keeping with the city's recent elevation to "the second-best place to raise children" in America by *Savvy* magazine (just behind Minneapolis-St Paul), or its rating as the most liveable city in America by both *Money* magazine, and *Places Rated Almanac*, the best-selling guide to living in the nation's towns and cities, a fact also broadcast on CBS news.[54] In addition, *Conde Nast Traveller* magazine rated the city "among the world's top ten travel destinations", whilst *Town and Country* proclaimed Seattle "the summer resort that has it all".[55] A rather eclectic range of publications, they nevertheless attested to the fact that when national periodicals chose to profile Seattle, they tended to be aimed at upscale consumers, and they were inclined to do so in order to foreground motifs of liveability.

In contrast to *American Heart*, *Sleepless in Seattle* was clearly designed to profit from the vogue for the city within mainstream popular culture in the early 1990s, and drew upon Seattle's reputation as a liveable city – a notion underlined by director Nora Ephron who stated in interview that Seattle had been selected because it was a city "where people have chosen lifestyle over work".[56] Indeed, its phenomenal success meant that the film became itself an important disseminator of images of the city as a liveable locale. However, it is important to note that the film also drew heavily upon a generic vision of the urban landscape that had much in common with romantic comedies set in other urban locations – a fact made clear by the filial Manhattan-set *You've Got Mail*. Thus whilst on the one hand the film mobilised the notion of Seattle as a unique urban locale, on the other it in fact relied upon the city's 'generic' liveable qualities. Indeed, as Kevin Robins points out, this is precisely the logic that underpinned the discourses of urban marketing and advertising of which liveability profiling was parasitic, and which sought out "distinction in a world beyond difference".[57]

SUBVERTING THE LIVEABLE: FROM STEPFATHERS TO SERIAL KILLERS

If *American Heart* suggested one approach to creating visions of family life in Seattle that sought to undercut the prevailing ethos of liveability, it was not the only available option. In the remainder of this chapter I want to examine two representations of Seattle that drew on the city's wider reputation for liveability whilst simultaneously undermining many of its core precepts, in the process generating macabre visions of terror within middle-class family life. The first is the modestly successful but highly acclaimed film *The Stepfather* (1987). Although it did not have a significant box-office impact upon release, *The Stepfather* can be used to illustrate the prevalence of liveability as a key aspect of Seattle's significatory repertoire in the period in question, and also suggest the range of ways the meanings of liveability were worked into different modes of fictional representation. Moreover, if *The Stepfather* was not a massive box-office success, it has nevertheless proved subsequently to have been an influential film, nowhere more clearly than in the way in which many of its key motifs were taken up within the second representation to be considered, namely the network television show *Millennium* (1996–99).[58]

The film director Joseph Ruben specialised in thrillers that offered sharp critiques of American middle class domestic relations. *The Good Son* (1993), written by novelist Ian McEwan, depicted a sociopath son (played by McCauley Culkin) wreaking havoc on his immediate family, whilst *Sleeping with the Enemy* (1991) charted the breakdown of a seemingly perfect marriage of an affluent young couple as the husband transformed into an abusive and brutal psychopath. The first of Ruben's trilogy of domestic thrillers was *The Stepfather,* a macabre thriller that drew much of its dramatic potency from the savvy way in which it played upon contemporary debates concerning the predicament of the nuclear family.

The beginning of the film depicts two of the central characters – Susan (Shelly Hack) and her teenage daughter Stephanie (Jill Shoelen) – fighting playfully in the back garden of their well-kept home in Seattle. As they continued to play-fight the mother joking shouts out "help! ... parental abuse!" This fleeting invocation of the parlance of contemporary family discourse served to indicate the movie's 'knowingness'; that it was quite self-consciously engaging with the welter of extant social commentary on the nuclear family. Indeed, the movie set up a well-worn theme of contemporary family dramas – Stephanie does not get along with her mother's new husband, Jerry Blake (Terry O'Quinn), much to the distress of Susan. In fact, much of the theme's efficacy came from precisely the fact that it was so quotidian; the fact that a child did not like a step-parent was so frequently the object of magazine articles, chat-show debates, movies, television series and self-help books that it had been normalised, and thus served as the perfect device for the thriller.

Stephanie's problems adjusting to life with her new stepfather lead her to be sent to a psychiatrist, Dr Bondurant, who attempts to find the cause for her newly developed pattern of disruptive behaviour in school. She tells the doctor her feeling that "if [Jerry] wasn't here, my mom and me would be all right". Dr Bondurant offers Stephanie a fairly standard textbook interpretation of her problems: that she is simply having difficulty adjusting to the death of her father and the introduction of her mother's new husband. The doctor opines that "you're going to have to face the fact that your mother loves the guy". This diagnosis and the likely prognosis are rejected by Stephanie, who, looking through the window at her stepfather waiting by the car, mutters something much more esoteric – protesting that her mother "doesn't *see*".

The irony is of course that Stephanie is right; that this is not simply the problem of a child adjusting to a new familial dynamic. The stepfather is a serial killer, who moves from one place to another attempting to create the perfect nuclear family. Thus when things inevitably 'go wrong' – in other words, when the everyday difficulties and disagreements of 'real' families occur, the stepfather slaughters the family, changes appearance and identity, and starts over again.

Stephanie's intuitive appeals might also have found a more sympathetic audience were it not for the fact that, to everyone else, Jerry seems the perfect father and husband. After buying Stephanie a puppy, to which the daughter reacts somewhat reservedly, Jerry says to Susan, "I hope she doesn't think I was trying to buy her love." Susan's reply is a comforting, "the puppy was perfect – *you're perfect*". Later on, Susan beseeches Stephanie to get along with Jerry with the claim that "he's a wonderful man, and he wants to care for us." Susan's words underscore the fact that she has been seduced by Jerry's simulacrum of benevolent patriarchy. Jerry is an all-singing, all-dancing, middle-class 1950s TV dad: fixing the swing door on the porch, building a bird table in his basement workshop, chuckling along to re-runs of *Mr Ed*. His language is formed from a nomenclature of chirpy up-beat phrases: he calls Stephanie "pumpkin" for example – that exists only in the nostalgic haze of *Father Knows Best* and *Leave it to Beaver*. As Stephanie confesses to a school friend:

> … he has this whole fantasy thing about how we should be like the families on TV … grin and laugh and be having fewer cavities all the time … I swear to God it's like having Ward Cleaver for a dad.

This is a vision Jerry himself is happy to endorse. Jerry sells real estate – or more precisely he sells perfect family homes to what he hopes are perfect families. At one point in the film he organises a barbecue in honour of the "first families I sold houses to". After having a photo taken of his family, Jerry proclaims that "I sell houses – that's my job, but it's more than that. Sometimes I truly believe that what I sell is the American dream." In a later scene, as a result of Stephanie's pronouncements Dr Bondurant becomes curious enough to undertake a covert attempt to meet Jerry. Disguised as a potential house-buyer, the doctor is shown around a property by Jerry, who states that "a house like this should really have a family in it". The doctor takes this as an opportunity to probe Jerry's psyche, and forwards that "you really are a cheerleader for the old traditional values, aren't you Jerry?" Jerry replies that "tradition is important". Probing further, the doctor opines that "it sounds like you had a strict upbringing". Jerry's suspicious and sinister reply is "you might say that".

Going down to the basement to his "all-American dad" work-shop, a deranged Jerry starts smashing and thrashing, and uttering that "all we need is a little order around here … keeping this family together … you had better believe it". Jerry, as the warped torchbearer for the 'traditional family', provides a demented vision of what could happen if Republican conservative family rhetoric was taken to its logical extreme. Yet as the scene with Dr Bondurant suggests, it is insinuated that Jerry has himself been the victim of parental abuse. Crucial to the movie's critique of conservative family values, it is implied that this abuse did not occur within the environment of a broken home, or as the fall-out from a latch-key existence, but rather, of a "strict" family upbringing. An upbringing that would, ironically, be regarded by right-wing commentators as a panacea for the moral degeneracy of family decline.

The coup de grace of the film's darkly satiric dismantling of 'family values' comes with the revelation of how Jerry chooses the areas in which to attempt to create his perfect families. The brother of a victim of Jerry's last 'failed' family has inherited the deathly shell of the house in which the stepfather went on his rampage. Going into the basement, he spots a magazine from which several pages have been ripped out. His curiosity leads him to the library, where he looks for a complete copy of the magazine. We learn there that the publication is *Travel and Leisure Magazine*, and that the missing pages are from an article entitled '10 great all American towns – ideal for raising a family'. Tempting as it is to conclude that the scene's intimation is simply that in order to take liveability profiles seriously, you (quite literally) have to be crazy, the implication is more wide ranging. By citing mainstream liveability profiles in this manner, *The Stepfather* served to indict the oeuvre, charging it as culpable in the instillation of a vision of family life as reactionary as it is fantastical. The stepfather, as the embodiment of the discordance between inner reality and the external projection of perfection, was the ideal target for the liveability paeans. In addition, the stepfather endorsed, albeit in the most extreme manner, many of the cherished motifs of liveability narratives – geographically mobile, the stepfather selected places to live on the basis of lifestyle criteria. Rather than an irrational aberration, the stepfather represented the apocalyptic end-point of a logical, valorised process of place consumption.

"A PLACE WHERE THEY CAN FEEL SAFE": MILLENNIUM

One of the most eagerly anticipated and high-profile television series of the mid-1990s was *Millennium* (1996–99), the creation of television writer and producer Chris Carter, who had also been responsible for the phenomenally successful sci-fi drama series *X-Files* (1993–2002). Largely on the basis of the success of *X-Files*, the American television network Fox gave *Millennium* a high-profile launch in its Autum 1996 schedule, the traditional time of year for networks to premiere their new shows. *Millennium* had been the subject of an exhaustive and lengthy promotional campaign, with teaser trailers and adverts in the trade press and in a range of sci-fi comics and publications. Perhaps most significantly, *Millennium* was given the *X-Files* slot in the Friday night prime-time Fox roster, a scheduling decision by the network that indicated a considerable commitment to generating a substantial audience for the show.

Millennium was not the ratings success that Fox had hoped, and failed to generate the momentum and cult fan base associated with *X-Files*. Faced with an increasingly dwindling audience, the show was cancelled before it could begin its fourth season in 1999. Whilst I will return to the question of *Millennium*'s 'failure' in due course, what I want to suggest is that despite the show's fairly swift decline and subsequent demise, it remains an important object of analysis. *Millennium* was conceived, marketed and scheduled as a major prime-time drama series, and it was set in Seattle. Moreover, the Seattle setting was a crucial aspect of *Millennium*'s identity, mobilised to generate a series of associations that infused the show's core thematic. In the following analysis I want to examine closely the pilot episode for the show. This by no means represents a comprehensive analysis of a show that lasted three seasons, and underwent some extensive revisions in the attempt to find 'an audience'. However, the pilot episode is of particular interest because it was the one that set out the key characters, ideas, places and premises for the show. It is thus an interesting example of how the creators and the network hoped that a major television drama set in Seattle would connect with a significant national audience base.

Millennium was an hour-long serious drama series that began with the return of Lieutenant Frank Black (Lance Henriksen) – a former FBI agent and criminal profiler attached to the bureau's serial-killer unit in Washington DC – to his hometown of Seattle. Frank, it almost goes without saying, is a family man. Early on in the Pilot episode, a police investigator asked Frank why he had come to Seattle. His reply was succinct and direct – "I'm here because I have a wife and a kid, and I want them to live in a place where they can feel safe."

Although *Millennium* was a work of narrative fiction, generically it depended on establishing a high level of social and cultural verisimilitude – indeed, it was the seeming spatial and temporal 'proximity' of the horrendous events generated in the show to those in the 'real world' that underpinned the logic of the drama. Thus the audience's initial acceptance of Frank's logic as 'sound' was predicated largely on the basis of the city's real-world role within the extant discourse of liveability – a discourse devoted to propounding the suitability of Seattle as a place for family life. In this way, Frank's statement had plausibility for the audience that such a declaration made similarly in defence of a move to Los Angeles or New York would simply not have possessed. [59]

Yet even before Frank chanted Seattle's liveable mantra, the episode had been working to undermine his words. The very first images provided by the episode are from a pre-credit sequence, supported by caption "downtown Seattle, February 2, 1996". Against a grey, rainy sky, the scene depicted a woman entering "Ruby Tip", a peep-show house, not dissimilar in look to the one depicted in *American Heart*. The camera scanned the interior of the establishment, picking up on excerpts of dialogue and images in order to construct the general milieu. It then focused on one dancer, and one peep-show watcher, and gradually revealed the man as a severely disturbed individual, uttering apocalyptic sounding axioms and imagining blood-soaked fantasies.

The scene was crucial to setting up the tension for the unfolding narrative of the episode, as Frank gets drawn into helping the local police track down this man, who turns out to be a psychopathic multiple murderer. Frank is compelled to use his 'gift' – one of the key dramatic devices of the show – namely the ability to see through the eyes of a killer, in order to apprehend the suspect. As Frank himself states, this gift is also a curse; a duality that echoes the dystopian and utopian visions of the Seattle that he also 'sees'. Frank's vision of the liveable city that he projected for his family was as fantastic as the apparitions of evil in his mind are horrific. As such, the peep-show house sequence served as an interesting allegory for the show's use of Seattle's liveable image. The peep-show house was similarly a site of duality – the screen the viewer looked through onto was the world as a site of projection and fantasy, of role play and wish fulfilment – the dancer asked – "tell me what you want." Yet the scene also showed us the mundane, real world of the dancers behind the scenes, complete with shopping bags, lockers, everyday items and the type of idle chit chat between co-workers that could just as easily take place in a shop or office.

Like the killer, Frank 'acted' on his fantasy, in the sense that he brought his family to a liveable Seattle. Thus the first scene after the title sequence could not be more different in look and feel to the preceding sequence just discussed. Frank is seen driving the family car; it is a bright, sunny day, and his wife Catherine (Megan Gallacher) and daughter Jordan (Brittany Tiplady) are beside him. They are moving up a quiet, well-kept residential street as Jordan looks excitedly out of the window, and Catherine covers her eyes, in anticipation of a surprise. The surprise is their new home in Seattle, "our new yellow house" as Jordan puts it, and Catherine's reaction is to say, "it's so beautiful". The home is an immaculate detached house, painted the sunniest and warmest of possible colours, complete with a welcoming porch, and depicted against the bluest and brightest of skies. It is also a picture that forms the final image of the title sequence, and is repeated throughout virtually every episode of the first two series. It is the symbol of the idealised, cherished family space that serves as the site of sanctuary and solace from the evil and darkness outside. Indeed, as the show's creator Chris Carter stated in interview, "for me the whole reason to do the show was that yellow house – a bright centre in a dark universe".[60] As they move to the house, Catherine states that "I think this move was the right thing", Frank replies, "I do too, it feels like home."

The scene continues, with Frank encountering a stranger outside his new house. The man hails Frank, introduces himself as "Jack Meredith" and states that "I guess we're going to be neighbours." Like Frank and his family Jack is white, but of retirement age, dressed smartly casual, and has a cheery, 'friendly-neighbour' demeanour. Jack is jovially inquisitive, and asks Frank where he is from and what he does for a living. To the later question, Frank replies that he "does some consulting". A wonderfully nebulous phrase, which serves to hide Frank's role as a member of the Millennium group, a secret "consortium of ex-law enforcement officers … formed to battle the darkness that approaches with the coming millennium".[61] Jack's reply is "ah, good" and asks Frank if "we can invite you folks over for dinner this week? – I see you have a little girl." Assured as Jack is now of the respectable, middle-class status of his neighbour – a standing enhanced by Frank's identity as a family man, the welcoming into the neighbourhood is complete. These subtle hints of status prejudice, and of neighbourhood parochialism – Frank has emitted the right signs to 'fit in' – makes the notion of what Jack would think if he realised what it meant to have "the Blacks" move into the neighbourhood more than a wicked verbal pun.

The scene ends with Jack and Frank parting, and Jack turning and adding the afterthought that "you couldn't have picked a nicer place to come back to" and offering a big 'thumbs up'. Frank smiles accordingly, and moves to read the local newspaper that has landed in his front yard. With Jack's psalm of liveability still reverberating, Frank scans the front cover before focusing in on a headline that reads "Mother found murdered at home. Five-year-old daughter hid from slayer." In that instant, all the fantasies of safety are destroyed, and the attack goes right to the centre of Frank's sanctuary – the family in the home, and his similarly vulnerable little girl. Frank turns around, and with a new sense of fear, thunderclouds are heard.

Whilst *Millennium*'s bleak world-view, esoteric aesthetic, and its obsession with mysterious government organisations, conspiracies and mayhem need to be situated in relation to its generic identity, and most specifically to its progenitor *X-Files*, its choice to situate that vision in Seattle was made precisely to emphasise the point that *nowhere* is safe – not even the nation's liveable city, valorised for its high ratings for bringing up the middle-class family, a point that it shared with its predecessor *The Stepfather*. However, unlike *The Stepfather*, *Millennium* did not critique the institution of the middle-class family – rather it used it as a motif of fragility in a precarious world. Arriving ten years later, *Millennium* was the product of a period which post-dated the Republican zeal for conservative 'family values', yet retained an investment in the talismanic status of the middle-class family.

Whilst it would be foolhardy to speculate on the complex issue of why *Millennium* failed to garner the audience that the network desired, it is worth noting in passing one of the strategies the network undertook as part of an ill-fated attempt to resuscitate the show in its final season. The decision was made to dispense with Frank's wife and child, relocate the show from Seattle and back to the FBI headquarters in Washington D.C., and to partner Frank with a female FBI agent in a manner akin to the male and female duo at the centre of *X-Files*. What was clear is the Seattle 'liveable setting city' and its attendant middle-class family was no longer deemed a required constituent of the show's core identity.

The range of texts examined in this chapter illustrate the fact that some of the most popular and critically acclaimed portrayals of Seattle generated during the period in question responded in complex ways to the discourse of liveability, and the ways in which it positioned the middle-class family in relation to the city. An understanding of the discourse, and a grasp of the prejudices and biases inherent within it, provides an important contextual framework for understanding how and why these representations emerged in the shape and form that they did, and when they did. These representations provided a range of mixed and often contradictory messages about Seattle's liveability, whilst some sought to contest outright its primacy as an interpretative framework.

CHAPTER 5

GRUNGE: 'A STORY THAT SOLD THE WORLD ON SEATTLE...'

Differentiated urban or local identities ... centred around the creation of an image, a fabricated and inauthentic identity, a false aura [are] usually achieved through 'the recuperation of "history" (real, imagined, or simply re-created as pastiche) and of "community"'.[1]

Your town is next... [2]

Grunge music, the term for a blend of punk rock and heavy metal played by some bands in Seattle from the late 1980s onwards, might seem an unlikely or incongruous subject area for a chapter in this book. After all, the preceding chapters have focused upon the subject areas of technology, nature, race and the middle-class family, all of which have been pivotal to the recent history of America's large urban centres. Grunge music, on the other hand, was a raucous musical idiom, specific to Seattle, and with few immediately perceivable connections to broader developments in American urbanity, or to the discourses that worked to mediate their comprehension. However, in this chapter I am concerned primarily not with the lyrics or melodies of grunge music – in other words, with the ways in which Seattle might have been represented through this specific musical idiom, though I necessarily touch upon this subject. I am interested centrally with the meanings of grunge as they were circulated by narratives disseminated through the national and international print media; that is, in the ways that grunge music was taken up as a subject for representation within a selected range of key texts from the period.

As this chapter will make clear, grunge music can be understood to have been implicated in two connected, but nevertheless discrete, discursive periods within American popular culture. The first period saw its representations restricted to the music press, one of the main organs for the dissemination of information, stories, reviews and profiles of popular music.[3] This period was characterised by an emphasis upon the mobilisation of Seattle's identity as the home for a 'unique' and 'authentic' musical idiom. As Stuart Hall points out, one of the features of the trend towards greater global interdependence has been an intensified interest in 'the local' and the particular.[4] As the discussion of 'Northwest lifestyle' in chapter two indicated, notions of Seattle's unique and 'authentic' identity were enunciated through the discourses of consumerism circulated within national and global media in the 1990s. Here, I want to draw out the ways in which such representations of Seattle need to be located in relation to specific historical trends within American popular music. As David Morley and Kevin Robins note, the desire to generate images and narratives of the distinctly 'local' for national and international publics has been closely linked to ways in which restructuring cities have found a key role for 'heritage' and 'tradition' in the creation of images and narratives of unique place identity.[5] Yet as M. Christine Boyer and Sharon Zukin point out, such practices are most usually associated with the mechanisms and logics of civic and entrepreneurial culture industries – urban civic elites working to impose new, tourist and enterprise-oriented visions of the city from 'above'.[6] What is apparent in the case of Seattle grunge, and what therefore makes it a quite intriguing example of a highly successful operation to mobilize notions of the homogenous 'local' for national and international publics, is that the process was instigated quite deliberately from 'below', in the form of the local record label Sub Pop.

The chapter then builds upon this necessary understanding of grunge's discursive origins to consider some of the ways in which it was taken up within a broader range of mainstream culture. In particular, I highlight the increasing awareness on the part of cultural institutions in the early 1990s regarding a lucrative and largely untapped market segment for cultural commodities, and quickly given the sobriquet of 'twentysomething' or 'generation x'. With close reference to a number of key accounts from the American print media from the time, I show how and why grunge music lent itself so readily to narratives of 'generation x'. Crucially, the chapter ends by outlining the peculiar appeal of Seattle grunge as a white urban musical idiom.

SETTING THE 'SCENE'

Grunge music had a profound impact on American popular culture in the early 1990s. Seattle grunge music bands such as Nirvana, Pearl Jam, Alice in Chains and Soundgarden were to be found high in the Billboard charts and on frequent rotation on radio and MTV. Band members were interviewed and profiled extensively and expansively in the regular news media and the music press. Ubiquitous retailers such as K-Mart and JC Penney emerged with stocks of the fashionable flannel shirts that signified the Seattle 'grunge look'. American network television screened adverts for products as diverse as Mountain Dew soft drink, AT&T call cards and Subaru Autos fronted by such aforementioned flannel-wearing, goatee-bearded 'grunge style' youths, hoping to appeal to the lucrative 18–25 demographic. What was clear was that grunge music, initially an imprecise term for a blend of punk rock and heavy metal being played by some bands in Seattle, had transcended its sonorial origins to become a full-blown media phenomenon. Moreover, if the look and sound of grunge seemed omnipresent, this did nothing to diminish the spotlight of attention focusing on Seattle, which became the signified epicentre of media hype.

In retrospect, the one moment that marked the instigation of the grunge music phenomenon was the release of the single *Smells Like Teen Spirit* by the Seattle band Nirvana in November 1991. Propelled by the incessant airing of the single's accompanying video on MTV, the band's album *Nevermind* was selling 400,000 copies a week by Christmas 1991. By January it had replaced Michael Jackson's *Dangerous* at the top of the Billboard album chart, and went on to gross $50 million within a year of release.[7] Clark Humphrey, whose book *Loser* remains the seminal cultural history of Seattle music, notes that "within weeks of *Nevermind*'s release, major-label scouts descended on alternative venues in Seattle (to find the 'Next Nirvana')".[8] What such activity implied was that Nirvana was understood not as a musical anomaly, but instead that Seattle was a seam rich with a precious commodity: a city with a unique and lucrative musical sound.

This was, of course, not a new notion. It was no accident that *Rolling Stone*, the US's most widely circulated and influential rock journalism magazine, declared Seattle in March 1992 'The New Liverpool'. The Mersey Beat era had defined the idea of city with a characteristic sound, complemented and consolidated by a vibrant local musical terrain. Similarly, Detroit in the 1960s (Motown), and New York and London in the 1970s (Punk/New Wave) were other cities cited frequently as fitting that profile. In addition, *Rolling Stone* identified Athens and Minneapolis in the 1980s as Seattle's most recent antecedents.[9] This seemingly smooth narrative of linear chronology connected this spatially disparate set of cities together as a pantheon of exciting music 'scenes'.

Whilst the phrase 'music scene' has long been a feature of journalistic parlance, academic writing on popular music in the 1990s attested to the fact that processes of globalisation, and their attendant impact on local, regional and community identities, made the notion of 'music scenes' a more intriguing subject of critical enquiry.[10] Writing in the journal *Cultural Studies* in October 1991, Will Straw cited Barry Shanks' use of the term 'scene' to account "for the relationship between different musical practices unfolding within a given geographic space".[11] The importance here was on the notion of difference, whereby a music 'scene' was a cultural space "in which a range of musical practices coexist, interacting with each other within a variety of processes of differentiation, and according to widely varying trajectories of change and cross-fertilisation".[12] Particularly pertinent was the fact that Shanks' definition of a music 'scene' reflected the growing influence of the concepts of space and nation within cultural theory, and was informing "the long-standing preoccupation of popular music scholars with the concept of community".[13] In contrast to Shanks' notion of a 'scene', the related concept of a music 'community' had traditionally been understood by music scholars as "a population group whose composition is relatively stable – according to a wide range of sociological variables – and whose involvement in music takes the form of an ongoing exploration of one or more musical idioms said to be rooted within a geographically specific historical heritage".[14] This was not to say that 'scenes', in Shanks' sense did not possess attributes that could most purposefully be understood in terms of 'community', such as engendering a (however transitory) sense of belonging, or of group identity. On the contrary, as Shanks pointed out, 'scenes' could be defined as "overproductive signifying communit[ies], a delineation that also served to liberate 'community' from its often hidebound signification".[15]

Yet at the same time as popular music scholarship was seeking to redefine the terms by which music 'community' and 'scene' might usefully be understood, representations of the grunge music phenomenon in the American music press and, a little later, in the general print media were performing a rather intriguing ellipsis. Despite some important distinctions between them – to which I will later return – both the music press and the mainstream media referred almost exclusively to a Seattle music 'scene', whilst the shape and form of what they were invariably depicting was much closer to the more traditional definition of a music 'community'.

My point here is not to berate non-academic discussions of popular music for failing to conform to a scholarly vocabulary. Rather, it is to note that whilst the range of popular music being produced and performed in Seattle in the late 1980s and early 1990s resembled most accurately Shanks' definition of a 'scene', in that it comprised successful hip hop, lightweight pop music, dance music, and rock-jazz hybrids, its *representation* in the national print media served to depict Seattle as a music 'community' in which the focus was upon the ongoing exploration of one musical idiom, namely grunge. Such representations placed a heavy emphasis upon notions of 'heritage', and 'tradition', supposedly rooted in a concrete and bounded place. For instance, Christopher Sandford, a music journalist who wrote extensively on Seattle music for *Rolling Stone*, provided a not untypical example:

> At the heart of the scene lay the factors underlying the Pacific Northwest that made it uniquely susceptible to the coming grunge revolution. [Seattle] has been called 'the hideout capital of the USA', a far-flung outpost of a town … Seattle is physically remote – closer to Russia than it is to New York … it is possible to think of the region *communally* … As early as the 1950s the area was synonymous with the raucous, guitar driven sound pioneered by the Sonics, the Wailers and the Kingsmen … the *tradition* of Northwest rock (emphasis added).[16]

What was being constructed here was a notion of Seattle as a locale with an organically grounded musical idiom, possessing detectable 'roots', and as, Sandford says, identifiable as part of a regional 'tradition'. Along similar lines, an article entitled 'A Seattle Slew' from *Rolling Stone* in September 1990 provided another typical instance of this sentiment. Noting that "Seattle, a city that until recently was famous for its rainy weather and inexpensive real estate … has become a regular music mecca", the article explained the city's appeal as stemming from the fact that "local groups … are unique, they're not in a major music market, so they don't follow any sort of formula at all".[17] Such claims for Seattle's (musical) insularity were also standard within biographical writing on Seattle bands in the late 1980s and early 1990s. For example, in his celebrated biography of Nirvana, Brad Morrell located Seattle as the latest arrival in a musical lineage stretching from the aforementioned Liverpool, London, and New York. Morrell evidenced more nuance in recognising that those cities were themselves music 'scenes', in Shanks' sense of the term, but added in contrast, "And Seattle? Its strength is its isolation."[18]

In asserting the 'strength' of Seattle's identity as a remote and 'closed' place, undisturbed by corrupting 'external' influences, narratives that constructed the city's insularity and isolation could be seen to have broad ideological implications. On the one hand, they could be seen to declare the uniqueness of Seattle as a location in the face of globalising forces that would seem to have worked to undermine old, supposedly fixed identities. Grunge, in this sense, became the signifier of what Kevin Robins terms, "a 'productive community' historically rooted in a particular place", a development in the valorisation of culture as "firmly rooted in well-bounded localities" that needs to be situated within the logics of global cultural economy.[19] As chapter two's discussion of the mobilising of images of 'Northwest Lifestyle' makes clear, related claims concerning Seattle also circulated in close association with other key texts that represented the city during the 1990s. Indeed, supporting the link between 'Northwest outdoors people' and the promulgation of a delimited racial profile for the city in the 1990s, it is important to note that grunge music was also portrayed as an exclusively white male pursuit – a notion to which this chapter will later return.

Yet the question of why Seattle was mobilised in such terms is complex, and should not be attributed simply to the abstract influence of 'global forces'. On the contrary, Seattle's portrayal as a homogenous musical community had a number of distinct discursive phases, the first of which limited itself predominantly to the national and international music press. A brief consideration of the genesis of the grunge phenomenon, and to the way in which it reveals an intriguing relationship not only to this strata of media, but also to the wider currents and trends in American popular music at this time, is necessary in order to situate and contextualise grunge's subsequent mobilisation within film and the general print media.

SUB POP:
A PACKAGE, AN IMAGE, AND A SOUND OF SEATTLE

In a piece for the Seattle weekly newspaper *The Stranger* in September 1993, Inga Moscio interviewed Jonathan Poneman, one of the co-owners and founders of the Seattle record label Sub Pop (the other being Bruce Pavitt). In response to what she suggested were grumbles of complaint in the city over the influence of the record label, Moscio prefaced her interview with the sardonic musing:

> Sub Pop did, arguably, put Seattle on the map of the United States. Sub Pop is responsible, albeit oftentimes indirectly, for a whole shitload of recognition our lovely city has fairly revelled in the past four years (sic).[20]

Although Moscio's article was not particularly influential in representational terms, it nevertheless alluded to the significance of Sub Pop in the marketing and dissemination of the Seattle 'sound', the Seattle 'scene', and in the process, key narratives regarding the city of Seattle. In this section I want to begin by considering in some detail the influence of Sub Pop in generating wider notions of Seattle as a discrete 'local' identity, with an organically grounded musical idiom, possessing detectable 'roots', and 'heritage', and identifiable as part of a regional 'tradition', before moving on to consider the appeal of such an identity more broadly.

Sub Pop began as *Subterranean Pop*, a fanzine "devoted to promoting independent records", started by Bruce Pavitt in Olympia, Washington in 1979.[21] By 1983, Pavitt had moved to Seattle, where Sub Pop became both a record review section in the city's music paper *The Rocket*, and a music show championing American independent record releases on local radio station

Figure 11: Hyping the label: cover of the Subpop 5 cassette, an early release for the company

KCMU.[22] By 1988, Sub Pop was a record company, co-owned by Pavitt and Jonathan Poneman.[23] Sub Pop's first releases were sampler compilations of local Seattle bands, entitled *Sub Pop 100* and *Sub Pop 200*. Jack Endino, an important local producer who was responsible for a significant percentage of the records released on the Sub Pop label from 1988 onwards, suggests that the label chose bands with a particular sound, described by Clark Humphrey as "dark and muddy".[24] Indeed, Humphrey notes that Nirvana's lead singer Kurt Cobain stated that he felt pressurised to tone down his "pop-songwriting sensibility" on *Bleach*, their first album, so as to "conform to a 'Sub Pop Sound'".[25] Moreover, that focus upon a particular sound was just one key element of Sub Pop's overall strategy, which was to foreground the label's own distinctive identity, rather than that of any particular band. As Chris Eckman, a member of local Seattle band The Walkabouts noted, Sub Pop's strategy was to "hype their label".[26] Poneman and Pavitt started the Sub Pop Singles Club, members of which "paid in advance for collectible 45s, most on color vinyl".[27] When combined with the limited release numbers, the Singles Club succeeded in generating underground demand and cachet for the label's output – to adapt a concept from sociologist Pierre Bourdieu, Sub Pop were intent on imbuing their records with *sub*cultural capital. As Endino notes, the result was that Sub Pop had a very desirable, marketable commodity – consisting of "a package, an image, and a sound".[28] According to Humphrey:

> [Poneman and Pavitt] devised a distinct image for the label: long hair, flailing heads, slam dancing and stage diving, beer, cynicism, male bonding, feedback, ear damage, smoke. No politics, no intellectualism, no fashion, no sincerity, no R&B, no women in sight … not only was this image an inaccurate portrayal of all Seattle bands, it was an inaccurate portrayal of all Sub Pop bands.[29]

Sub Pop's activities thus saw them deliberately distilling a multi-sonorous music scene into a marketable 'sound', emanating from a homogeneous 'community'. As Humphrey notes, "it sold, particularly to the image-hungry British music press".[30] Ironically, given the discursive constructions of Seattle as the site of an isolated music community, one of the most important aspects about grunge music's emergence served to emphasise the flows of information, influence and capital across local, national and international space. This was apparent from the fact that from early on, the promotion of grunge music was dependent on the currents of trans-Atlantic musical influence; favourable coverage in the British music press, in the form of profiles in *NME*, *Sounds* and, most importantly, the high-circulation *Melody Maker*, provided "the best way to gain credibility in the American underground market".[31] Poneman and Pavitt seemingly understood this, and in early 1989 paid for *Melody Maker* journalist Everett True to fly to Seattle and report on the new 'Seattle scene'.

On 11 March 1989, *Melody Maker* ran a cover article on Seattle bands, entitled 'Sub Pop, Sub Normal, Subversion', written by True, and cited in retrospect as a key article in the instigation of grunge music as a musical and marketing phenomenon. The article stated that:

> Britain is currently held in thrall by a rock explosion emanating from one small, insignificant, West Coast American City. Seattle … has a new claim to fame

"SUB/POP 5 is to subterranean American music what WANNA BUY A BRIDGE? was to the once underground Brit scene. A blast of fresh air and conclusive proof there's life in the land yet." —NEW YORK ROCKER

"SUB/POP is the best index there is of American local independent scenes." —NEW MUSIC EXPRESS

Well hi there: S/POP now alternates, quarterly, between a C-60 trans-regional cassette and a networking newsletter. S/POP 5 is available on GENERIC tape ($4.00 ppd.) or TDK normal bias ($5.00 ppd.) S/POP 6 is free (send an SASE). S/POP 7 out May 15, is available on TDK only ($5.00 ppd.). Please write immediately to the LOST MUSIC NETWORK, BOX 2391, OLYMPIA, WA 98507.

U.S. UNDERGROUND

Figure 11: Hyping the label: cover of the Subpop 5 cassette, an early release for the company

KCMU.[22] By 1988, Sub Pop was a record company, co-owned by Pavitt and Jonathan Poneman.[23] Sub Pop's first releases were sampler compilations of local Seattle bands, entitled *Sub Pop 100* and *Sub Pop 200*. Jack Endino, an important local producer who was responsible for a significant percentage of the records released on the Sub Pop label from 1988 onwards, suggests that the label chose bands with a particular sound, described by Clark Humphrey as "dark and muddy".[24] Indeed, Humphrey notes that Nirvana's lead singer Kurt Cobain stated that he felt pressurised to tone down his "pop-songwriting sensibility" on *Bleach*, their first album, so as to "conform to a 'Sub Pop Sound'".[25] Moreover, that focus upon a particular sound was just one key element of Sub Pop's overall strategy, which was to foreground the label's own distinctive identity, rather than that of any particular band. As Chris Eckman, a member of local Seattle band The Walkabouts noted, Sub Pop's strategy was to "hype their label".[26] Poneman and Pavitt started the Sub Pop Singles Club, members of which "paid in advance for collectible 45s, most on color vinyl".[27] When combined with the limited release numbers, the Singles Club succeeded in generating underground demand and cachet for the label's output – to adapt a concept from sociologist Pierre Bourdieu, Sub Pop were intent on imbuing their records with *sub*cultural capital. As Endino notes, the result was that Sub Pop had a very desirable, marketable commodity – consisting of "a package, an image, and a sound".[28] According to Humphrey:

> [Poneman and Pavitt] devised a distinct image for the label: long hair, flailing heads, slam dancing and stage diving, beer, cynicism, male bonding, feedback, ear damage, smoke. No politics, no intellectualism, no fashion, no sincerity, no R&B, no women in sight ... not only was this image an inaccurate portrayal of all Seattle bands, it was an inaccurate portrayal of all Sub Pop bands.[29]

Sub Pop's activities thus saw them deliberately distilling a multi-sonorous music scene into a marketable 'sound', emanating from a homogeneous 'community'. As Humphrey notes, "it sold, particularly to the image-hungry British music press".[30] Ironically, given the discursive constructions of Seattle as the site of an isolated music community, one of the most important aspects about grunge music's emergence served to emphasise the flows of information, influence and capital across local, national and international space. This was apparent from the fact that from early on, the promotion of grunge music was dependent on the currents of trans-Atlantic musical influence; favourable coverage in the British music press, in the form of profiles in *NME*, *Sounds* and, most importantly, the high-circulation *Melody Maker*, provided "the best way to gain credibility in the American underground market".[31] Poneman and Pavitt seemingly understood this, and in early 1989 paid for *Melody Maker* journalist Everett True to fly to Seattle and report on the new 'Seattle scene'.

On 11 March 1989, *Melody Maker* ran a cover article on Seattle bands, entitled 'Sub Pop, Sub Normal, Subversion', written by True, and cited in retrospect as a key article in the instigation of grunge music as a musical and marketing phenomenon. The article stated that:

> Britain is currently held in thrall by a rock explosion emanating from one small, insignificant, West Coast American City. Seattle ... has a new claim to fame

– the Sub Pop recording emporium … It's the city's very remoteness (2,000 miles from Chicago) that has been a major cause of Sub Pop's distinctive sound. Freed of the constraints of peer pressure that bands receive on the East Coast, Seattle bands have been given time to develop, away from the harsh demands of fashion.[32]

Journalistic hyperbole concerning Britain's 'thrall' aside, the article worked to establish, for a national and international public, the importance of Seattle's 'remoteness', and of its 'backwater' status. Even the choice to note Seattle's distance from Chicago, rather than the more obvious cities of Los Angeles or San Francisco was carefully selected in order to construct Seattle's spatial peripherality and insularity. This in turn helped in the depiction of the music as a discovery; a hidden secret, allowed to develop untainted by interference from corrupting market forces. Such markers of 'purity' and 'distinctiveness' acted to signify the legitimacy of the music within the taste culture of the American underground music scene.[33]

What was central to the genesis and appeal of the narrative of the Seattle music 'scene' that Sub Pop was instrumental in constructing was the notion of an individual identity and sound. As *Melody Maker* made clear, Seattle did not represent innovation in the sense of technological development within the music – these were "raw distorted guitar bands".[34] Rather, what was being implied was that Seattle offered what Will Straw has termed a "regional, authorial vision", and in 1989, that gave the city and its 'sound' a major status within rock music discourse.[35] As Straw argues, the traditional, predominant narrative of American rock music's history, as it has been established through rock journalism, literature and promotional discourses, has been one of an "ongoing succession" of regional, authorial visions. For example, Straw discussed the successful marketing of "the so-called US 'heartland'" rock associated with artists such as Tom Petty or John Cougar Mellencamp in the mid-1980s.[36] The problem was that by the late 1980s, this narrative, always questionable empirically, was becoming unsustainable. Reflected in the falling sales for such artists, rock was understood by the industry to be in crisis. Moreover, at a discursive level, nationally circulated, 'mainstream', influential organs of rock journalism such as *Rolling Stone* were instrumental in depicting other popular musical idioms such as heavy metal, R'n'B, and dance music in the ascendancy. Straw quoted a *Billboard* interview with Hugo Burnham, the A & R director at the major record label Island Records in 1990, in which Burnham suggested that "Rock music is losing ground because there's not an awful lot new that's happened to rock music since punk."[37] It was into that void that Sub Pop, aided and abetted by the eager British music tabloids, was able to "promote a handful of bands playing in a few tiny bars in a far-off port city as the potential saviours of rock".[38]

The notion of rock requiring a saviour was on one level, related to specific developments in the alternative music scene in the 1980s. As Straw has argued, by the early or mid-1980s "a terrain of musical activity commonly described as 'alternative' was a feature of virtually all US and Canadian urban centres".[39] This situation had come about as the result of events occurring in the time subsequent to the punk music explosion. As Straw notes:

> As local punk scenes stabilised, they developed the infrastructures (record labels, performance venues, lines of communication, etc.) within which a variety of other musical practices unfolded. These practices, most often involving the eclectic revival and transformation of older musical forms, collectively fell under the sign of the term 'alternative'.[40]

Crucial to an analysis of the culture of alternative rock is the place within it for the notion of musical progression and of "regional, authorial vision". In terms of musical progression, Straw argues convincingly that "it was no longer the case, as it has been in the period immediately following punk, that change would involve the regular displacement of styles as the historical resonance of each emerged and faded".[41] What occurred was temporal stabilisation, rather than the divesting of out-moded elements. In terms of "regional, authorial vision", such a notion was seen as increasingly obsolete, as "each local space has evolved, to varying degrees, the range of musical vernaculars emergent within others, and the global culture of alternative rock music is one in which localism has been reproduced, in relatively uniform ways, on a continental and international level".[42]

Within such a context, it is easy to see why the vision of grunge music emanating from Seattle would prove so seductive and enticing. Articles such as those written by Everett True for *Melody Maker* consolidated Sub Pop's attempt to generate the sense of the city as possessing a unique, locally-specific musical sound. The narrative pronounced that this was not simply another interchangeable alternative music terrain, but an explosive, primal rock'n'roll, allowed to foment in geographic isolation, and now ready to take over the world. In short, grunge promised the return to the notion of a regional, authorial vision for American rock.[43]

GRUNGE GOES MAINSTREAM: ALL THE RAGE, NONE OF THE GUILT

In his writing on the British punk scene, Dick Hebdige uses the term 'recuperation' to refer to the way in which subcultural signs, such as music and dress are converted into commodity form, using the example of the fashion spreads on Punk in the British *Observer* newspaper.[44] Yet as he notes, "the relationship between the spectacular subculture and the various industries which service and exploit it is notoriously ambiguous … It is therefore difficult to maintain any absolute distinction between commercial exploitation on the one hand and creativity/originality on the other".[45] The key role played by the independent record label Sub Pop in distilling Seattle's multi-sonorous music scene into a marketable 'sound', emanating from a homogeneous 'community', provided a good example of the symbiosis of creativity with processes of production, publicity and packaging – what Hebdige calls "artisan capitalism".

During 1992 and 1993, there emerged a range of products and marketing campaigns in the mainstream media which seemingly attested to the attraction of grunge's recuperation into commodity culture. Memorable examples included a 1993 advertising campaign for the Subaru Impreza, in which a 'grunge'-clad youth sneered "this car is like punk rock", against the background noise of heavy guitar music. The American phone company 1-800-Collect ran a commercial that featured an elderly woman dressed in the long johns, flannel shirt, and stocking cap associated with grunge music, and making a phone call to Seattle. The national print media featured adverts for grunge-inspired physical fitness classes entitled "grunge aerobics", and discount clothing stores advertised their extensive range of childrens' "grunge-wear". [46] By far the most infamous single instance of grunge's recuperation

Figure 12: Grunge goes overground: Pearl Jam's Eddie Vedder on the cover of *Time* magazine

SELLING SEATTLE

into 'profitable merchandise', and one that echoed the *Observer* colour spreads on punk cited by Hebdige, was American *Vogue*'s grunge fashion spread in December 1992.[47] Explaining that grunge "had broken out of the clubs, garages and thrift shops of Seattle to dominate rock radio, MTV and the aspirations of kids across America", the magazine devoted ten pages to a piece entitled 'Grunge and Glory', in which models wearing highly priced 'grunge haute couture' by Ralph Lauren and Calvin Klein were photographed by top fashion photographer Stephen Meisel.[48] A particularly noteworthy moment in the *Vogue* fashion spread was a small caption which accompanied one of the female models, and stated "make-up is minimal to non-existent, with only the mouth occasionally richly coloured, *à la* Courtney Love".[49] The appearance of such a caption suggested the presence of a fair degree of irony in *Vogue*'s fashion spread, which in Hebdige's terms represents another strategy by which mainstream culture "diffused and defused" subcultural style.[50]

Nowhere were the complexities of grunge's incorporation into mainstream culture more apparent than in the production of *Singles* (1992), the first and only Hollywood film to represent Seattle as the home of grunge music.[51] Written and directed by Cameron Crowe, *Singles* was a romantic comedy with an ensemble cast of white, middle-class, young adults all living in the same apartment building in Seattle. Crowe, a former journalist for *Rolling Stone*, stated in interview that *Singles* was conceived as a story about "disconnected single people forming their own loose family", and living in an apartment building that functioned as "the halfway house between college and getting your own place".[52] The director explained that he had envisaged *Singles* originally as a story set in Phoenix, Arizona, but that he moved the location to Seattle when he "discovered the budding Seattle music scene".[53] The filming of *Singles* began in Seattle in March 1991, whilst the film's eventual release was in September 1992.[54] The life of *Singles* therefore straddled the period in which Nirvana's phenomenally successful album *Nevermind* was released. Crowe's recounting of his initial approach to Warner Bros., who would subsequently finance the film, evidenced just how dissimilar those two periods were. Referring to his choice to locate the film in Seattle, the director stated that "the studio was … less than thrilled with [the] chosen setting, viewing it as some cold, rainy city up north".[55] Crowe's reaction was to tell "these guys … it *has* to be in Seattle. The air is different. The music's different, and they were like 'yeah, yeah,

yeah'."[56] Interviewed in September 1992, Crowe noted sardonically that "now of course, these same people are going 'Is Nirvana in the movie?'"[57] Crowe's remark regarding his desire to shift the location of *Singles* from Phoenix to Seattle due to "the budding Seattle music scene" thus had a quite different set of implications in September 1992 than it did in March 1991. In particular, the intervening period made it appear that the film had been produced in order to profit from the subsequent heavy demand for Seattle-signified grunge commodities.[58]

Singles was a reasonable box-office success, no doubt aided by the high profile of Seattle grunge bands. Yet the film itself has to be understood as existing in a quite complex relationship to these other relatively non-narrativised representations of grunge. *Singles* was a romantic comedy that weaved interrelated story lines, connecting the cast in platonic and romantic relationships. Although the main characters were all depicted as having some relationship to the Seattle grunge scene, one even as a grunge band member, most were only shown as occasional fans frequenting gigs downtown, the 'grunge scene' functioned more as the affective backdrop, as the best-selling grunge soundtrack marked shifts in the film's mood and tone. Visually, the film mapped the space of the city in quite interesting ways. It began with a series of seemingly random shots of Seattle, although in actual fact all were connected thematically by their signifiers of vibrant, 'hip' city life. An intriguing filmic comparison would be the beginning of Woody Allen's *Manhattan*, which maps the city with a series of discrete shots of monumental architecture. By contrast, *Singles* used the same cinematic device for mapping the city, but instead chose intimate tableaux, in the process reflecting the smaller scale of the actual city, but more importantly, serving the function of giving the city a small scale 'community' feel, which was reflected in the film as a whole. The suggestion of homage to Woody Allen was also apparent in the film's use of *Annie Hall*-style character monologues to camera, which also had the effect of orienting the audience towards its central theme of intimate dialogues between characters. The film ends with an elaborate and ostentatious shot, which zooms out from the apartment complex in which all the characters resides, cranes upwards, and then pans across the entire cityscape. In one sense, this shot reflects the fact that the loose fragments with which the film started had been tied together, as the characters' narrative strands had been resolved. It also has the effect of uniting the characters and the city, and underscored as the shot is by a babble of chattering voices, this scene of closure implies that the city is teeming with such 'singles' style connections.

If the links to other romantic comedies implied that the generation of meaning in *Singles* was more complex than any suggestion of a straightforward representation of the 'real' Seattle 'scene', it is also true that the film needed to be situated in relation to wider developments in contemporary culture. For instance, *Singles* was released within months of the Fox network television drama series *Melrose Place* (1992–99), itself soon followed by the NBC sitcom *Friends* (1993–2004), both of which, generic distinctions aside, were highly successful fictional narratives of relatively affluent young adults engaged in communal urban living. In this way, *Singles* should also be understood within the context of a more general move to tailor aspects of popular culture towards a lucrative and recently discovered 'twentysomething' market. What is clear is that the commercial success of grunge coincided with a more wide-ranging fixation on the part of the American media with comprehending the youthful demographic that comprised grunge's artists and fans.

TIME FOR GENERATION X

From the beginning of the 1990s, mainstream magazines exhibited a marked concern with identifying and describing young Americans between eighteen and thirty years of age. For instance, in July 1990 *Time* magazine regarded the search for this demographic significant enough to make it the cover story, in the process creating something of a prototype for such articles. Entitled 'Proceeding With Caution', the *Time* story stated its intention to explore and explain the values of a group it termed "the twentysomething generation".[59] The article stated that

> they are an unsung generation, hardly recognised as a social force or even noticed at all ... by and large, the 18-to-29 group scornfully rejects the habits and values of the baby boomers, viewing that group as self-centred, fickle and impractical ... a prime characteristic of today's young adults is their desire to avoid risk, pain and rapid change.[60]

Central to the *Time* article, and to a corresponding range of profiles in upscale national print publications such as the *Atlantic Monthly* ('The New Generation Gap'); *Business Week* ('Move over Boomers'); and *U.S. News & World Report* ('Just Fix It!') was that 'the twentysomethings' were in opposition, and often in conflict, with the values of the baby-boomer generation that had preceded and raised it.[61] Echoing Todd Gitlin's comment about the baby boomer's shift from "'J'accuse' to Jacuzzi", the articles focused upon 1990s twentysomethings as 'damaged' by the previous generation's transformation from "twentysomething hippies to thirtysomething yuppies".[62] The baby boomer's liberal attitudes to parenting, marriage – "an estimated 40 per cent of people in their twenties are the kids of divorce" and subsequent 1980s acquisitiveness – "flashy toys and new clothes were supposed to make up for this lack" had produced a "worried", "tolerant" generation "that could help repair the excesses of rampant individualism".[63] Considering the upscale 'boomer' demographic of the range of publications in which these articles appeared, the twentysomething narratives comprised something of a pseudo-ethnographic study of the younger generation.

If such pronouncements read like abstracted generalisations, it may be because they were, but the power and predominance of this discourse could not be denied. Nor could the overwhelming need to create this discourse be overlooked; these articles evidenced an obsession on the part of the mainstream media with identifying and characterising this cohort of young Americans. What made this process so attractive, and for *Time* provided the avowed principal motivation, was the supposedly enigmatic character of the 'twentysomethings' – "they have few heroes, no anthems, no style to call their own" combined with the fact that the cohort was about to "begin their prime working years".[64] Crucially, then, it was no accident that the obsession with understanding the composition of the 'twentysomethings' coincided with a growing awareness of its reputation as a cohort with significant and largely untapped disposable income.

Thus at the same time as mainstream periodicals ran articles on the 'twentysomethings', "the baby busters" or, after the title of Douglas Coupland's successful and influential novel, "Generation X", American advertising publications were providing guides for companies on how to promote products to the new primary target market.[65] For instance, in

Figure 13: Matt Dillon (far right) and the band members of Pearl Jam in Cameron Crowe's *Singles*

August 1993 *American Demographics* promoted a new publication entitled *The Baby Bust: A Generation Comes of Age*, a marketing tool hailed as "the first statistical biography of this important generation", and full of ostensibly essential information on the tastes and opinions of the twentysomethings. Preceding *American Demographics* by only a few months, the February 1993 edition of *Advertising Age* ran a news article entitled 'The Media Wakes Up To Generation X', warning marketers that "twentysomethings are $125B market, but a lot different from their boomer forebears".[66] Aimed at an advertising industry readership, what made this article particularly noteworthy was that it provided an analysis of the strategies then being undertaken by magazines such as *Rolling Stone* and *Time*:

> As they begin to consciously pursue Xers, print and electronic media companies are following two paths. Some are trying to appeal to both boomers and busters … The *Time Inc.* weekly is reformulating sales and marketing strategies to reach … both generations. Recent covers of … *Rolling Stone* … once considered the boomers' bible … have featured a mix of youth oriented 'grunge' bands like Nirvana.[67]

Anticipating generational narcissism, one of prime motives for the plethora of articles on the twentysomethings in mainstream periodicals and weeklies was thus the hope of garnering increased readership from that very demographic, which would in turn make them more attractive to advertisers. What is also apparent is that one of the best prospects for gaining the attention and the dollars of the twentysomethings was understood to be the strategy of foregrounding grunge-related stories.

From a financial point of view, the use of grunge-related stories in national print publications from 1991 onwards made perfect sense. As has been stated, Nirvana's *Nevermind* had made the band the biggest-selling act in popular music, but they were by no means the only signified Seattle act to achieve explosive sales and publicity. [68] In Nirvana's wake came Pearl Jam, whose debut album *Ten* had been released almost simultaneously with *Nevermind.* Although *Ten* took longer to accrue sales, an intensive touring schedule and the impact of the single (and accompanying video for) "Jeremy" (appropriately, a tale of damaged childhood) helped it to sell 4 million copies in the US alone by the end of 1992. In addition, other grunge record releases, such as Soundgarden's *Badmotorfinger*, Alice In Chains' *Dirt*, Temple of the Dog, the *Singles* soundtrack and the reissued Mother Love Bone CD all found themselves amongst the top 100 selling albums of 1992. [69] Thus the popularity of grunge, and the apparent appetite for its appeal, made it a reasoned choice of topic for publications eager to attract a bigger slice of its purchasing demographic.

Yet the appeal of grunge music to these publications was more complex than mere revenue-chasing. Whilst advertising dollars might explain the decision to foreground grunge related stories, it was not enough to account for the actual type of stories these publications generated. For example, apparent was the fact that grunge music was easily assimilable to existing generic modes of narrative carried by publications such as *Time.* In particular, grunge provided a conventional and attractive narrative framework by which to articulate particular developments in American youth culture. For instance, consider the contrasting paradigm suggested by dance music culture, also influential in the 1990s. The logics of dance music, particularly in the early 1990s, were distinguished by a general disregard for many of the things that rock held sacrosanct, and which grunge so easily incorporated. [70] In contrast to rock, dance music did not valorise the individual 'artist' and the 'artist's' career as the emblem of progress and development; the creator of a particular dance track might have been unnamed, unknown, but most usually irrelevant. Moreover, the dance track was not identified as the vessel for communicating the "artist's" inner truth, nor was there an assumed privileged place for the album, rock's talismanic symbol of a coherent, authorial vision. Thus if the mainstream press were looking for the conventional narrative focal point provided by a "voice of a generation", they were not going to find it in dance music culture, which was, if anything, an example the fragmentation of youth culture, rather than its unifying cure. Grunge, in contrast was more assimilable to that remit.

Arguably the most striking example of the capacity of grunge to lend itself to this form of discursive application was provided by the cover of the 25 October 1993 issue of *Time*. This cover, and the feature article that accompanied it, can be understood as a key moment in the incorporation of grunge music into mainstream commercial culture. On the front cover was a photograph of Eddie Vedder, the lead singer of the Seattle grunge band Pearl Jam, then one of the best-selling bands in the United States. The torso and face of Vedder covered the whole of the cover of *Time*, as the singer was portrayed grasping a microphone and standing against the backdrop of a stage. Vedder had been photographed 'mid-scream', and his face was an image of focused intensity with the suggestion of emotive pain. The photograph was accompanied by a large heading that ran 'All The Rage', while the text underneath explained that: "Angry young rockers like PEARL JAM give voice to the passions and fears of a generation." Leaving the accompanying text momentarily to one side, what is clear is that grunge music was able to fulfil one of *Time* magazine's signature mandates, namely to embody newsworthy social and cultural developments in the figure of one emblematic individual. Moreover, Vedder's place on the cover of *Time* conferred him entry into the magazine's pantheon of popular music figures. Indeed, the text of the cover story went further, stating that:

> They haven't built that Rock 'n' Roll Hall of Fame in Cleveland, Ohio, yet, but when they do, they'd better save a room for Vedder. He's got all the rock-idol moves down. Does he have a painful, shadowy past? Check. Does he have an air of danger and sensuality reminiscent of Jim Morrison? You bet.[71]

In connecting Eddie Vedder, by association, with previous 'rock-idols', *Time* was constructing grunge music's canonical heritage, and thus indicating a range of previous meanings and associations through which this new phenomena could be interpreted, appraised and assimilated. Indeed, the most important reason for the inclusion of Pearl Jam, or more specifically Eddie Vedder (since what was also being foregrounded here was the authorial vision of the singer-songwriter) in such a prominent *Time* story was the suggestion that the band represented the "voice … of a generation".[72] As the article stated "alternative rock is the sound of homes breaking … an emotional soundtrack, speaking directly to unresolved issues of abandonment and unfairness".[73] When considered alongside *Time*'s earlier, pre-grunge article on "the twentysomethings", these words suggest that for the magazine, alternative rock had provided the freshly minted soundtrack for a generation stated previously to have had "no anthems".[74]

What was also pivotal to *Time*'s representation of Pearl Jam and Eddie Vedder was that they were part of a musical idiom that could be firmly established in a specific geographical space. As the article stated, Seattle was "a seminal music scene … a grunge mecca".[75] Again, the contrast with dance music is instructive. Highly polycentric, and characterised by the simultaneous existence of large numbers of local or regional styles, dance music and the identity of the musical loci was in constant mutation and fluctuation. In terms of the urban dance music audience, there was a marked heterogeneity, exemplified in the constant criss-crossing of the boundaries of race, gender and sexuality, which itself reflected the incorporation of diverse influences into the music; female soul voices, for example, or the origins of dance music within gay club culture.[76] For a publication such as *Time*, grunge's capacity to provide more traditional, established, and thus graspable genealogies of temporal and spatial musical development made it easily assimilable to its general readership.

Yet the real point of contrast for grunge music, and arguably the key to much of its seductiveness for a range of mainstream press, was with hip hop. Hip hop, an overwhelmingly black musical idiom, first emerged in America's large urban centres in the 1970s. During the late 1970s and early 1980s, it remained very much a low profile musical idiom, a street culture with a market "still based inside New York's black and Hispanic communities".[77] However, by the early 1990s, things had changed significantly, so much so that hip hop had become "one of the most heavily traded popular commodities on the market".[78] As Tricia Rose points out, the identity of hip hop was inseparable from its urban context, namely "life on the margins of post-industrial urban America".[79] Elaborating on the crucial role of the post-industrial city on the shape and direction of hip hop, Rose states:

> The explosion of local and national cable programming of music videos spread hip hop dance steps, clothing and slang across the country faster than bushfire. Within a decade, Los Angeles County (especially Compton), Oakland, Detroit, Chicago, Houston, Atlanta, Miami, Newark and Trenton, Roxbury and Philadelphia [had] … link[ed] various regional post-industrial urban experiences of alienation, unemployment, police harassment and social and economic isolation to their specific experience via hip hop's language, style and attitude.[80]

What is clear is that hip hop represented a vibrant, incendiary, fundamentally urban musical idiom, with an aesthetic, political and cultural resonance which went straight to the heart of the contemporary urban condition. Moreover, if, at the end of the 1990s, hip hop had become relatively depoliticised, this was most assuredly not the case in the late 1980s and early 1990s. High-profile artists such as Public Enemy associated themselves closely with the teachings of Malcolm X and Black Power, as well as the Nation of Islam. Moreover as Murray Foreman points out, this period of hip hop was also characterised by the ascendancy of the "specific discursive construct of 'the 'hood'".[81] As mobilised by best-selling rap artists such as Snoop Dogg, Ice T, Niggaz With Attitude (N.W.A), Dr Dre and the late Tupac Shakur, the 'hood was crucial to the dominant "themes, images and postures that [took] the form of the pimp, hustler, gambler and gangster", and the overarching narratives of "ghetto poverty and gang aggression".[82] Spurred on by a series of highly publicised shootings between West Coast (Los Angeles) and East Coast (New York) hip hop crews, 'Gansgsta Rap', mediated images and discourses of gang culture, the black underclass, urban poverty, drug and gun culture for audiences, (often white and suburban) outside of the 'hood.[83]

As both Rose and Foreman make clear, to represent hip hop, particularly at this specific historical juncture, was inevitably to draw attention to the extreme, racialised schisms in the urban fabric, the detrimental impact of post-industrial restructuring upon non-white urban populations, the chronic problems of homelessness, poverty, unemployment, drug dependency, violence, hopelessness and structural inequalities which found their spatial fix in the nation's large urban centres.[84] This was the case even if, paradoxically, such post-industrial urban context was simultaneously being aestheticised, fetishised and commodified for suburban consumers.

However, it is not necessary to deny the attraction of hip hop to still make a strong case for the symbolic appeal of grunge. What grunge offered was a distinctively urban musical idiom, with all the signifiers of tradition, locality and authenticity, characterised by vibrancy and a potent and lucrative mixture of youthful anger and ennui, talismanic yet largely apolitical individuals, without the unmistakable overtones of post-industrial decay and racialised inequality closely associated cities such as Los Angeles and New York. As a white urban musical idiom tied to notions of 'generation x', grunge offered the mainstream media 'all the rage' with none of the guilt. This was clear in the way that *Time* magazine, by 1993, was able to draw upon other key aspects of Seattle's predominant signifying repertoire, alluding to the arrival of grunge with the statement that "sometime in 1991, Seattle became more than a quintessentially livable city where the coffee was strong, the people were friendly and the plastic was recycled".[85] Clearly, there is much more to be said about grunge music, and I do not want to suggest that this chapter represents a comprehensive coverage of the complex connections between the musical phenomenon and Seattle. What I have sought to do is to draw upon a number of key representations of grunge in order to situate its appeal within the context of larger theoretical issues pertaining to American urbanity in the late 1980s and 1990s. In particular, what I have tried to draw out is the nuances of the mobilisation of notions of Seattle as a homogenous, discrete music 'scene', and to show how they can be situated in relation to two discrete periods of the city's signification. The importance of Seattle in signifying a 'well bounded locality', with a unique musical idiom served a particular function for the American music press in the late 1980s, struggling to cope with the difficulties posed by the global culture of alternative rock. The incorporation of grunge into mainstream narratives of 'generation x' in the early 1990s drew upon many of those signifiers, reshaping them to present needs, in particular utilising their ability to be easily assimilated to existing generic modes of narrative carried by publications such as *Vogue* and *Time*. In particular, grunge provided a conventional and attractive narrative framework by which to articulate particular developments in American youth culture, and offered conventional and stable notions of musical progression. Perhaps most crucial was the ability of grunge, via Seattle, to offer the chimera a distinctively white urban musical idiom – to offer 'all the rage' with none of the guilt carried by the bleak, decaying post-industrial landscape of hip hop. However, it is important to recognise that the close association between the city and a 'branded' consumption practice such as grunge also attested to the profound shift in urban economies in the last twenty years, and it is to another key element of this, namely the prolific rise of Seattle-signified 'gourmet coffee', that we now turn.

CHAPTER 6

'PLANET STARBUCKS':
SELLING SEATTLE GLOBALLY

Almost any product which has some tie to place – no matter how 'invented' this may be – can be sold as embodying that place.[1]

During the 1990s gourmet coffee, available in the first instance for purchase and drinking in specialist coffee shops in American cities, emerged as a phenomenally successful new consumption practice. The primary initiator, and by far and away the largest and most important of these specialist gourmet coffee retailers, was the Seattle company Starbucks. The proliferation of Starbucks and other gourmet coffee retailers at this time was no accident; indeed, it needs to be understood in the context of the profound socio-spatial restructuring of the nation's urban centres over the last twenty years. Sharon Zukin provides an exemplary rendering of these developments when she states:

> With the disappearance of local manufacturing industries and periodic crises in government and finance, culture is more and more the business of cities – the basis of their tourist attractions and their unique competitive edge. The growth of cultural consumption (of art, food, fashion, music, tourism) and the industries that cater to it fuels the city's symbolic economy, its visible ability to produce symbols and space.[2]

What marks out recent times is thus the magnitude, intensity and ubiquity of such consumption practices in the context of transformed urban centres experiencing the myriad forces of a "local-global political economy".[3] As Stuart Hall remarks, "globalisation, in the form of flexible specialisation and 'niche' marketing, actually exploits local differentiation".[4] Central to the organisation of such consumption practices is therefore the mobilising of images and narratives of place. It is not only that cities have sought to mobilise images and narratives of their 'unique' sites of cultural consumption within what are understood as increasingly vital promotional and marketing strategies. As the discussion of Eddie Bauer and REI in chapter three, and Sub Pop in chapter five makes clear, it is also the case that companies offering new consumption practices have made close associations with specific places fundamental to their advertising and branding activities. In doing so, such companies have also been involved in the process of mediating the meanings of those places for the specific purpose of attracting potential customers. They have done so in a manner that connects them in complex ways to other modes and registers of the city's representational repertoire, and also to the material form of the urban environment.

In theorising the complex "intertwining of cultural symbols and entrepreneurial capital" at the heart of the contemporary urban economy, Zukin provides a useful frame for this chapter's focus on the new consumption practice of gourmet coffee.[5] One of the key ways in which images and narratives of Seattle were generated in the 1990s was through the promotional and design strategies employed by Starbucks, and other successful gourmet coffee retailers such as Seattle Coffee Company, and Seattle's Best Coffee, and enunciated on-site, within the built environment of the city. Such companies elaborated their connections with Seattle quite explicitly within a repertoire of promotional materials, and also associated themselves with a range of more oblique symbols of urban liveability and Northwest lifestyle already extant and explored in this study. Considered alongside the ways in which the notion of Seattle as a capital of 'coffee culture' gathered momentum and widespread dissemination on film, television and the print media, such materials represented a major contribution to the city's heightened profile in the 1990s. In the following section I situate the emergence of gourmet coffee retailing in relation to wider shifts within American urban culture and transformations in the global economy, before going on to look more closely at how and why Seattle was mobilised in the selling of gourmet coffee.

THE GROUNDS FOR SEATTLE COFFEE

Food has proven a particularly useful object of analysis for cultural theorists seeking to understand the role of urban spaces and places within a globalising economy. As David Bell and Gill Valentine point out:

> Culinary cultures constructed as 'original', 'authentic' and place-bound … can be deconstructed as mere moments in ongoing processes of incorporation, reworking and redefinition: food is always on the move, and always has been. [6]

Nowhere has this notion of 'food on the move' been made more apparent than in the history of coffee. A food commodity that brings with it a long and complex chronicle of consumption and production, coffee is a perennially global product par excellence. As Michael D. Smith suggests, "the history of coffee is indefeasibly bound up with the rise of capitalism and the overseas expansion of European colonialism".[7] By the early nineteenth century plantations had been set up in the colonial territories of Central and South America and Africa usually "relying on some form of slave or indentured labor, to serve the growing popular demand in an industrialising and increasingly urbanised Europe".[8] Indeed, the consumption of coffee was central to the historical development of urban European culture. As Margaret Visser notes, in England the coffee house "became associated with work as well as leisure, a place where men could drink coffee, read newspapers … and transact some kind of business". [9] The most famous example was Lloyds coffee house in London, "where insurers met, [and where] the organisation of Lloyds 'names' remains to the present day".[10] The nineteenth-century coffee house also provided the cheapest eating place "for the growing middle classes working in the cities", and its numbers swelled throughout the century in order to "meet the demands of the 'mercantile age'".[11] Coffee houses also had a reputation on the continent as sites for the bourgeoisie to engage in "literary, artistic and political gestation".[12] As Paul Andrews notes, coffee houses have been cited with some frequency as the places where the "seeds of the 1848 and Russian revolutions were planted, [and] psychoanalysis was given early debate", whilst the Parisian cafes "provided a fertile garden for [the] literary and artistic geniuses of the past century".[13] Thus apparent is the fact that coffee's role in the development of mercantile capitalism, bourgeois urban culture and the global exchange of capital and commodities is longstanding. What is more, as Smith points out, at the end of the twentieth century coffee continued to be "the most important tropical commodity in the international agricultural trade", whilst also retaining its longstanding "relationship of dependency that connect[ed] the poor, underdeveloped nations of the South to the rich, industrialised countries of the North."[14]

Whilst the history of coffee suggests a range of specifiable connections between the commodity and a range of notable spaces and places, it nevertheless also indicates the difficulty of fixing it to any one of them – coffee, to repeat, has always been 'on the move'. Yet the notion of gourmet coffee – a 'niche' market exploiting its connection to the local – can be situated much more specifically in place and time. Gourmet coffee, with its allusions to food connoisseurship, is intimately tied to notions of taste, which as Pierre Bourdieu argues, operate to generate distinctions between people, most customarily by social class.[15] It is thus no coincidence that the very first outlet for the sale of gourmet coffee was the first Starbucks shop in Seattle's Pike Place Market, opened in 1971.[16] Pike Place Market, a downtown open-air food market, was one of Seattle's first forays into the type of "revitalisation" identified by Neil Smith and Peter Williams as central to the processes of contemporary urban restructuring across the United States.[17] Part of a new phase of urban development that represented "the primacy of consumption over production", Pike Place Market was an instance of the refashioning of urban space to the taste and aesthetic of the expanding class of white-collar workers in the financial and business service sectors.[18] Like many other urban 'historic preservation' sites across the United States emerging at this time, the desire to 'keep' Pike Place Market, and thus retain a sense of 'the local', could be understood in terms of what Erica Carter, James Donald and Judith Squires identify as "paradoxically both a resistance to, and at the same time a product of … global forces".[19] In this sense, Pike Place Market, a typical instance of Zukin's burgeoning urban symbolic economy, was therefore an entirely appropriate site for the genesis of gourmet coffee.[20]

Yet if the milieu in which gourmet coffee emerged as a 'niche' market explains much, it still does not indicate why it should occur in Seattle – after all, the city's urban restructuring was indicative of developments occurring in urban centres across the United States. However, the desire to embark upon a more extensive search at the level of the material form of Seattle as an urban environment – in other words, searching for what makes the actual city 'distinct' – is somewhat missing the point. As Bell and Valentine point out, what is crucial here is the mobilising of distinctiveness at a discursive level; marketing a product as 'place-specific' for the purpose of profit, a process often requiring little more than an 'invented' tie to 'the local'. This fixing of 'place-specific' distinctiveness at the level of association – what Bell and Valentine call 'commodity biographies' – has the capacity to shift the position of commodities within what Bourdieu terms "the economy of cultural goods", creating distinctions between homogenised, mass-marketed products (such as Nescafe powdered coffee) and those identified with the unique associations of 'place'.[21]

Much is revealed by a brief consideration of the chronological trajectory of Starbucks. The company was bought by Howard Schultz in 1987, and it was under his ownership that it was transformed from being a local seller of whole-bean gourmet coffee to becoming an international chain of specialist coffee shops.[22] Upon purchase Schultz announced his plan to "position Starbucks coffee as the first national brand of speciality coffee". Schultz was also explicit in outlining the philosophy behind his strategy of expansion, stating at the time that his "primary interest is in Chicago, because Chicago has a high degree of sophistication, a high downtown density and a good economy". Schultz added that he was confident that his plan would work "because outside of Seattle and San Francisco, there are few gourmet coffee roasters or successful espresso bar businesses".[23] His confidence was well founded – at the time of takeover in 1987, there were 17 stores in the Seattle area, by 1994, there were 425 across the United States, including outlets in New York, Los Angeles, Washington, D.C., Denver, Boston and Atlanta.[24] Schultz's statement outlining his plan to situate coffee shops in "busy financial districts in other cities" announced the sort of clientele he was intent on attracting, namely the urban professional middle classes, working in financial service industries and related fields.[25]

In this way, Schultz's plans for expansion into urban financial districts corresponded with a period in the late 1980s that, as Robert Beauregard notes, witnessed a marked increase in investment in central business districts and the expansion of corporate and retail services.[26] Gourmet coffee, as a 'niche' commodity, was a consumable valorised for its exchange value as much as its use value, and its supposed ability to bestow cachet and status. Rather than being taken home to be drunk, consumption was on display, in chic coffee bars, the perfect consumable commodity for the advanced service city, giving capitalism its caffeine boost, whilst at the same time helping to keep coffee's stock buoyant on the world market. As Michael D. Smith notes, "Starbucks outlets are integrally connected to those 'landscapes of leisure' where people with disposable income, not to mention cultural capital, go to consume, display themselves, and watch others." [27] Indeed, Smith notes that Starbucks innovated by inverting the urban architectural logic epitomised by Mike Davis' term "kill the street". Davis uses the term to refer to the way in which Los Angeles' urban restructuring in the 1980s worked to construct upscale business, retail and residential architecture away from the public space of the city, thus functioning as a prime manifestation of that city's 'spatial apartheid'.[28] By contrast, Starbucks utilised large windows, and a significant percentage of "out-facing" seating, thus inviting the customer to "eat the street".[29] There are therefore parallels to be made with complementary developments in urban socio-spatial restructuring outlined by M. Christine Boyer in her study of the revitalisation of New York's South Street Seaport from disused waterfront zone to high-class shopping and tourist centre. Boyer views the function of what she terms such "city tableaux [as] the places where a society turns back upon itself to view the spectacle of its own performance".[30] It is arguable that Starbucks has been a key exponent in the spatial and ideological diffusion of this practice. The company has expanded its stores outwards from carefully demarcated "revitalisation" zones in America's urban centres and towards a more variegated selection of city spaces, and increasingly, towards spaces of consumption and leisure in the shape of suburban shopping malls both nationally and internationally. Central to that expansion has been the continued mobilisation of specific images of Seattle by gourmet coffee retailers, and it is to the details of that process that we now turn.

THINK SEATTLE, ACT GLOBALLY

In March 1998 a futuristic indoor shopping centre opened at Cribbs Causeway, a retail park outside the city of Bristol, England. Containing over 130 shops, the eponymously named Mall's archetypal layout combined two wings with large department stores at the tips, and a central core offering a number of food outlets on multiple levels. [31] A considerable section of the central core had been allotted to the Seattle Coffee Company, a speciality coffee bean and beverage retailer. The Seattle Coffee Company had been started by Seattleite Ally Svenson, who had relocated to London, and opened her first store in Covent Garden (a key site of London's urban renewal) in 1995.[32] Svenson had added another 55 outlets by the time the company was acquired by Starbucks in the middle of 1998.[33]

At the Seattle Coffee Company, store customers could choose to take out, or consume inside, espresso, cappuccino or caffé latte drinks, amongst a myriad of other coffee-based beverages. Alternatively, they could purchase a bag of gourmet whole bean coffee to grind at home, specifying their orders for 'Mount Rainier Blend', 'Pioneer Square Blend' or 'Seattle City Blend'. With the aid of a plethora of promotional signage and leafleting, in the hub of a shopping centre in the southwest of England customers were being encouraged to "take part in THE experience" of Seattle. [34]

The word 'Seattle' functioned ostensibly as shorthand for the Seattle Coffee Company, as in the store pamphlet's declaration that "here at Seattle we believe in a 'coffee experience'". The pamphlet's encouragement to customers to "call to find the closest Seattle to your work or home" was a manifestation of the company's overall strategy to propose the synonymy between the company, the city and gourmet coffee, as was the suggestion that the customer call to "just have a chat about Seattle".[35] What was being enunciated here was the notion of Seattle Coffee as the "iconic product of a particular place".[36] Indeed, the Seattle Coffee Company's pre-eminent trademark icon was a silhouette of the Seattle city skyline, which featured on the company's promotional literature, coffee cups, store fronts and fascia. This iconic marketing strategy not only allowed the company to foreground the consumption practice's origins in a specifiable geographic 'place', but also worked to suggest the consumption of gourmet coffee as a symbolically 'urban' and 'urbane' activity, and thus draw upon cosmopolitan associations within its actual ex-urban shopping mall location.[37]

This mobilisation of "the closest Seattle" was symptomatic of the reproduction of 'the local' within what Nigel Thrift terms the "distinctive economic, social and cultural systems" of global consumerism in recent times.[38] Indeed, by the end of the 1990s, the experience of "the closest Seattle" could quite rightly claim to be a worldwide phenomenon, stretching from the Seattle Coffee Shop in Cape Town, South Africa, to Starbucks Coffee International locations in Taiwan, Thailand, New Zealand and Malaysia.[39] This reflected the status of speciality coffee retailers as one of the leading purveyors of new consumption practices proliferating in urban and ex-urban locations around the globe.[40]

As stated previously, Starbucks was by far the market leader in speciality coffee retailing, with over 2,500 stores in North America, the UK, the Pacific Rim and the Middle East by the end of the decade, and with net revenues of $1.7 billion in 1999.[41] Although the ways in which Starbucks mobilised its association with Seattle was not as brazenly iconographic as the Seattle Coffee Company it acquired, it was nevertheless a core element of its complex signifying repertoire. Underlining the importance of the articulation of 'unique' place identity, Starbucks' advertising literature stated that one of the company's "defining characteristics" is "where we came from", and the company made no secret of its Seattle origins; its promotional literature, packaging, and the naming of some of its coffee blends all portrayed its Pacific Northwest 'origins'.[42] As Smith points out, "Starbucks has cultivated an image as a Pacific Northwest phenomenon", and it is clear that its image fitted closely with the other elements of 'Northwest Lifestyle' identified in chapter three, as well as constructions of the city's liveability, and this is something that will be returned to later.[43] Indeed, the more oblique design features of Starbucks stores supported a particular rendering of 'urban life' which fitted closely with other key elements of Seattle's significatory repertoire. As Smith notes, décor was tasteful 'earthy green and brown, brassy trim, and lots of glass' with the privileging of a range of design motifs and features that were intended to connote 'authenticity' and 'quality'.[44] As such, Starbucks combined elements of urban liveability with signifiers of nature and naturalness which were in keeping with a mainstream discourse of Seattle.

REPRESENTING SEATTLE'S 'COFFEE CULTURE': PRINT PRESS, TELEVISION AND FILM

Prior to 1990, the American mainstream print media made no mention of coffee amongst the variable repertoire of signifiers used to depict the city.[45]

Arguably the first article to do so was written by Mary Schmich for the *Chicago Tribune* in March 1990. A brief 'lifestyle' profile outlining Seattle's elevation in the charts of urban liveability, Schmich pointed to its current identity as "the sex-symbol of cities". Yet Schmich argued that if Seattle was a sex symbol, it was a "PG-13 kind", and added "it's a Marilyn Monroe who sips cappuccino in one of the ubiquitous downtown espresso bars", thus serving to promulgate notions of the city's genteel and civilised urban identity.[46] In addition, Mary Bruno's article for *Lear's* in July 1991 mentioned in passing the impact of Starbucks in creating "a coffee culture that is unparalleled in any other American city".[47] By December 1992, Ann Japenga, in her article on 'Northwest Style' for the *Los Angeles Times*, noted that Starbucks was "bringing Seattle-style coffee culture to cities including Chicago, Denver, San Francisco and San Diego".[48] It is also worth noting that from the early 1990s onwards, tourist guide books to Seattle started including sections devoted to paying homage to the city's "coffee culture". For instance, a representative sample would include *Rough Guide*, which stated that "Seattle [has] turned coffee consumption into an art form"; *Lonely Planet,* which asked its readers if they had "ever wondered whether caffeine is a viable substitute for sunshine? If so, Seattle is your kind of town"; *Fodors,* which featured a section with the title 'Welcome to the Home of the Coffee Break'; and *Insight Guides: Seattle*, which contained a whole chapter entitled 'Coffee and Culture', noting that "people drink coffee here like nowhere else in America".[49] Such tourist publications, existing in a mutually advantageous and supportive relationship with urban symbolic economies, thus had a considerable interest in amplifying such 'unique' signifiers of cultural consumption.[50]

One of the most illuminating illustrations of the pervasiveness of the association between Seattle and gourmet coffee by the mid-1990s is provided by the NBC prime-time sitcom *Frasier* (1994–) discussed in chapter three. A highly successful and critically lauded show, *Frasier* was one of the most high-profile and long-standing representations of Seattle within American popular culture. As noted earlier, *Frasier* was a 'spin-off' from the award winning prime-time sitcom *Cheers* (1982–93), and relocated one of *Cheers*'s large cast of regular characters, psychiatrist Dr Frasier Crane (Kelsey Grammer) from his original location in Boston's *Cheers* bar to his 'hometown' of Seattle. Yet there had been no mention of Frasier's 'hometown' during his time as a character in *Cheers*. According to *Frasier*'s creator Peter Casey, the team working on the original script design for the sitcom had location 'carte blanche'. As Casey stated in interview, the choice of Seattle was made due to the fact that "the whole coffee revolution was getting publicised, coming out of Seattle, and so that began to really appeal to us".[51]

Whether Casey's words were an accurate reflection of the key factors in the choice of *Frasier*'s location or not, what was clear is that the sitcom included as one of its three key sets the gourmet coffee shop 'Café Nervosa'. The café was the frequent location of the main characters' romantic pick-ups and verbal put-downs, witty retorts and the social *faux pas*, and where the gazes from strangers and the cases of mistaken identity regularly set up the storylines of particular episodes. Moreover, for aesthetes as ardent as the show's Frasier Crane and his brother Niles, the refined surroundings of Café Nervosa and the cultural distinctions of a cappuccino or a caffé latte provided the slender, but nevertheless crucial, markers of taste and refinement. In this way, 'Café Nervosa' provided the apposite *mise-en-scène* for an instance of 'quality television' meant to attract a 'blue chip', upscale, well-educated, urban-dwelling demographic.

Of course, it is important to recognise that *Frasier* did not simply 'reflect' an extant Seattle 'coffee culture', but worked influentially to mediate perceptions and knowledge of it through the generic form of the 'quality' sitcom. As the bar setting in *Frasier*'s progenitor *Cheers* made clear, the 'quality' sitcom format lent itself to the foregrounding of a regular site of 'public' consumption, and the opportunities for character interaction that it provided. The movement towards a coffee shop setting in *Frasier*, and its Manhattan-set counterpart *Friends* (1994–) thus not only reflected the attempt on the part of both shows to generate an appositely 'fashionable' urban setting for their audiences, but also exemplified the way in which, as Steve Neale and Frank Krutnik point out, the sitcom format usually structures itself around a small number of regular settings.[52]

The depiction of Seattle-style 'coffee culture' within fictional representational forms in the 1990s needs also to be linked more closely to the particular promotional strategies of companies such as Starbucks. As Naomi Klein points out, "brand-name product placement in films has become an indispensable marketing device for companies like Nike, Macintosh and Starbucks".[53] For example, Starbucks paid to have its products placed in the aforementioned urban romantic comedy *You've Got Mail* (1998) and the US prime-time 'quality television' show *Ally McBeal* (1998–2002), a romantic comedy drama series set in a fictional Boston law firm. Both *You've Got Mail* and *Ally McBeal* were savvy sites for place promotion, in the sense that they generated appositely upscale, stylish urban milieu for a desirable brand association.

However, two other high profile instances of Starbucks's brand-name product placement seemed to fit less smoothly with the requisite 'coffee-culture' images, namely *Fight Club* (1999) and *Austin Powers: The Spy Who Shagged Me* (1999). These two films by no means represented the 'tip of an iceberg' of extensive product placement by Starbucks. Indeed, the choice to foreground them here is an easy one – together with *Ally McBeal* and *You've Got Mail*, they comprised virtually the sum total of the selective televisual and filmic promotional campaign undertaken by Starbucks in the 1990s. As Klein notes, like the Body Shop, Starbucks represented one of a number of 1990s brand retailers who largely eschewed large-scale advertising campaigns in favour of applying carefully tailored, niche marketing techniques.[54]

Fight Club, a critically acclaimed and successful Hollywood film adapted from the novel by Chuck Palanhiuk, portrayed the self-destructive path of a disaffected, ennui-ridden white-collar worker. A bleak action-thriller, *Fight Club* showed its central character suffering from a dose of existential angst familiar from a range of popular cultural texts from *The Catcher in the Rye* onwards, albeit with a decidedly contemporising hue. A central theme in the film was the character's dejection with and rejection of consumer culture, which propelled his 'descent' into the netherworld of underground 'fight clubs', full of young men seeking physical catharsis from 'numbing' everyday life. An early scene in the film depicts the chain of events catalysing the main character's descent, as he rants about a consumption-oriented, therapy-group fixated *gesellschaft*. As he narrates his internal monologue of weary dejection, he rebukes a world carved out from instantaneous Ikea interiors, and other suggested artefacts of pseudo-individuation. Crucially, the name he gave to this world was 'Planet Starbucks'.[55] A sardonic spin on actual 1990s corporate entities such as 'Planet Hollywood' and 'Planet Reebok', the moniker conveyed the character's uneasy sense of the totalising, corporate lifestyle narrative he felt embodied and disseminated by the Seattle-based speciality coffee retailer.[56]

The scene was fairly brief, and it seems that the company's choice to have itself portrayed in such a light suggests that it felt that the benefits of product placement in front of *Fight Club*'s audience demographic would outweigh any lasting impact of the brand's specific signifying function within the diegesis, which was, at the very least, ambiguous, coming as it did from the mind of a disturbed individual.

An even more intriguing example of Starbucks's active investment in what seemed ostensibly to be 'negative imaging' was the Hollywood comedy *Austin Powers: The Spy Who Shagged Me*, one of the highest grossing films of 1999. A sequel, *Austin Powers* was the second outing for the eponymous secret agent, a figure created as a pastiche of 1960s debonair playboy screen spies such as James Bond and Derek Flint.[57] As a pastiche spy film, *Austin Powers* also had an arch-villain, in the shape of Dr Evil, a parody of James Bond's nemesis Ernst Blofeld in *You Only Live Twice* (1967). The film begins with Dr Evil being rescued from outer space, where he had been free floating since the late 1960s. Brought to earth with a bump in 1999, Dr Evil is informed that his cohort of accomplices have been busy at work maintaining the infrastructure of the 'evil empire' in his absence. To such an end, they have set up a new headquarters for the development of plans for world domination, at the top of the Space Needle in Seattle. This knowledge is accompanied by an establishing shot that depicts the Space Needle against a dark stormy sky, complete with foreboding music, as lightning strikes the Needle's mast. Emblazoned around the Space Needle's revolving restaurant, shining green and menacingly against the night sky, was the word 'Starbucks'.

Crucially, *Austin Powers* supports the argument that Starbucks' association with Seattle was a key element of the company's promotional strategy, despite the fact that the company's outlets had become an omnipresent feature of urban centres throughout the United States, and could thus be mobilised by the Boston-set *Ally McBeal*, or the Manhattan-set *You've Got Mail*. Indeed, the humour of the Starbucks scene in *Austin Powers* depends upon the fact that audience knowledge of the link between the city and the brand pre-existed the viewing of the film. More to the point, the humour of the scene relies upon the audience having a *particular kind* of knowledge about Starbucks – the kind that would make the company's portrayal in the film as the heart of the 'evil empire' meaningful.

Austin Powers' portrait of Starbucks as the 'evil empire' thus drew humorously upon a set of negative associations that had become attached to the company as the 1990s progressed. Reflecting the film's identity as a comedy, the depiction of Starbucks as 'evil empire' was couched in humour, and the villains of *Austin Powers* were eminently likeable as they plotted global domination whilst supping their Starbucks coffees. Indeed, it could be argued that the attraction of the film for Starbucks lay in the way in which such a high profile opportunity for self-mockery enabled it to defuse and co-opt recent criticism of its behaviour, re-working the meanings of such criticism within a fictionalised and suitably 'witty' representational vehicle.

The criticism that Starbucks had received in the 1990s occurred on two fronts. The first was to do with what Klein called the company's 'clustering expansion methods'.[58] Starbucks had attracted some high profile negative publicity due to its alleged strategy of dropping "clusters of outlets in urban areas already dotted with cafés and espresso bars".[59] For instance, a well-publicised case in the San Francisco suburb of Mill Valley in 1996 saw locals "accuse … Starbucks of trying to bump off their local coffee house".[60] In relation to the Mill Valley episode, *Business Ethics Magazine* carried a quote from Matt Patsky, an industry analyst, who stated that it "resulted in a lot of bad publicity for Starbucks. They were depicted as the evil empire."[61]

The second front on which Starbucks was criticised also related to the notion of it as an 'evil empire' – this time attached much more closely to the phrase's intimation of lingering post-colonial power-relations, and in particular, to the company's treatment of its coffee growers in Guatemala. As Bell and Valentine note, the 1990s witnessed a "shift in the politics of food consumption" as a range of books disseminating knowledge on 'world food problems' jostled for shelf space with 'round the world cookbooks'.[62] In particular they suggest that there was an increased unease…

> [a]bout the role export pressure of exotic produce plays in sustaining and even deepening inequalities in new global relations of capital accumulation dominated by transnational corporations.[63]

As they point out, such concerns had worked to re-orient consumption 'towards ethical positions'.[64] A key exemplar of this was the growth in products such as 'fair trade' tea and coffee, which by foregrounding their production sources, and issues of workers' wages, education and working conditions, "reveal[ed], rather than mask[ed] the links in the commodity chain".[65] From 1994 onwards Starbucks became the target for activists from the Coalition for Justice for Coffee Workers (CJCW) and the US/Guatemala Labor Education Project (GLEP). CJCW and GLEP activists began holding rallies and leafleting many of Starbucks' stores as part of an attempt to get the company to institute a code of conduct "to ensure that the plantation workers who pick the coffee

Figure 14: A protestor exits a Starbucks shop via the broken window during the WTO demonstrations in Seattle

are paid a living wage and have decent working conditions".[66] Starbucks CEO Howard Schultz's reaction was to state that Starbucks "embraced many of the things that [the CJCW] believe", and followed the lead of other American corporations such as Levi Straus and Reebok by instigating a code of conduct in February 1995.[67]

The targeting of Starbucks drew much efficacy from the knowledge that the company had taken great care to present itself as progressive and socially responsible. Indeed, as Naomi Klein points out, brand-based companies such as Starbucks were particularly appealing to campaigners, not only because such companies' iconic status gave them a celebrity that made negative revelations 'newsworthy', but also because they relied so much upon the power of their brand image for success in the marketplace.[68] Starbucks has traded upon the company's commitment to "contributing positively to our communities and our environment, underlining the importance of such activity to its brand image".[69] As Schultz himself admitted in reference to the attempts of companies to project a social conscience, "trust is the future of brands".[70] That 'trust' was increasingly perceived to be open to scrutiny in a world characterised by "the rapidly developing and ever denser network of interconnection and interdependencies" – what John Tomlinson calls the "complex connectivity" of globalisation.[71] Intriguingly, Starbucks released an official statement in 1994, in response to the activities of GLEP and the CJCW, insisting that since they represented "less than 0.5 per cent of the total coffee market … the perception is that Starbucks Coffee can make more of an impact than what is truly possible".[72] Yet, as the strategy of the activist groups suggested, and Starbucks' acquiescence admitted, that perception was precisely the result of the company's own zeal to "build an enduring global brand" that had less to do with the company's share of the raw commodities market than with its ostensible ability to exude significant cultural power and influence.[73]

Interestingly, Starbucks' own promotional strategies had a quite complex relationship to its production spaces. As Bell and Valentine point out, the "creation of geographical knowledges or 'lores' about commodities" is a complex process that can involve a "double commodity fetish" that incorporates "certain ignorances … together with certain place constructions or knowledges".[74] The promotion of 'place' within the strategies of selling gourmet coffee exemplified this complexity. As noted earlier, it might be assumed that coffee's links to complex post-colonial power relations might be something that gourmet coffee retailers would work hard to conceal from their customers. Yet as Michael Smith points out, Starbucks was keen to foreground "narrativised geographies and evocations of exotic, far-off realms" in its literature.[75] Starbucks has produced what it calls a 'World Coffee Tour card', rewarding the customer who 'travels' "from the exotic islands of Indonesia to the misty peaks of the Andes" with Starbucks coffee.[76] The result of such promotional material was to stir up "images of exotic Third World locales that persist as a sort of colonial sediment in the popular imaginary of the West".[77] Contrary to Fredric Jameson's assertion of "the effacement of the traces of production" in the processes of contemporary consumption, Smith notes that gourmet coffee retail has involved "deliberately emphasis[ing] the relations of production and mak[ing] them part of the commodity itself".[78] Smith calls this process "coffee tourism", yet he himself retains a quite conventional conception of "the relations of production", and uses the term to refer only to the countries in which the coffee beans are shown to be cultivated. Yet as this chapter has shown, Starbucks generated a much more complex range of place constructions, in which its role as the "tour operator" was at least as much about the promotion of a journey to an "imaginary Seattle" as it was to an "imaginary exotic Third World". Illuminating in this respect were the words of Giles Whittell in the London *Times*, who stated that for him, gourmet coffee provided "a way to have a piece of the northwestern dream life wherever you [happen] to live".[79]

SPILT COFFEE: STARBUCKS AND THE WTO

If one moment could be said to exemplify both the success and failure of the complex strategies of place promotion undertaken by the gourmet coffee retailers within a "local-global political economy", it was Seattle's hosting of the World Trade Organisation (WTO) Ministerial Conference Meeting in Seattle between 30 November and 3 December 1999. As the Office of the United States Trade Representative's press release stated, the meeting was intended to "launch global negotiations to further open markets in goods, services and agricultural trade".[80] Yet the highly publicised protests and

Figure 15: *Newsweek*'s spectacular cover typified the mainstream press's coverage of the events in Seattle

demonstrations that accompanied, and largely overshadowed, the reporting of the WTO Ministerial Conference in both the national and international media served to suggest that there was more than one way to interpret the significance and intention of proceedings.

Accounts in the press stressed repeatedly the eclectic political and ethical grievances and objections of the demonstrators. The estimated 40,000 people on the streets of Seattle included 20,000 members of the ALF–CIO (American Labor Federation–Congress of Industrial Organisations), whose organised union march against the effects of free trade on American workers received relatively little press coverage; NGOs (non-governmental organisations) such as Greenpeace, protesting against the proliferation of genetically modified organisms in food and agriculture; and groups such as Ralph Nader's Public Citizen (multinational corporations), the Sierra Club (rainforest logging), Confederation Paysanne (hormone-fed beef), n30 (global capitalism) and the headline-catching Sea Turtle Restoration Group (turtle-killing shrimp nets).[81] Indeed, *Newsweek*'s Michael Elliott summed up the irony of this international network of protestors (many making their influence felt via cyber 'sit-ins' powered by Microsoft browsers) with the musing that "there are now two visions of globalisation on offer, one led by commerce, one by social activism".[82]

The demonstrations suggested that the WTO's vision of globalisation, so clearly and unproblematically laid out in their press release, had been called into question in a myriad of (often conflicting and incompatible) ways. On the 30 November Paul Schell, the mayor of Seattle, called in the National Guard, who with the aid of pepper sprays, tear gas, rubber pellets, riot staves and concussion grenades "established a 'protest-free' zone encompassing the convention centre and major hotels".[83] As the mayor declared a civil emergency and placed most of downtown Seattle under curfew, most national and international newspapers carried photos of the police in "RoboCop-like riot pads and helmets" restraining the protestors against hazy tear-gassed streetscapes.[84]

Todd Gitlin, writing in *Newsweek*, summarised events pithily with the remark that "as with all public eruptions, symbols soaked up the attention".[85] Ironically, Gitlin's observation could have been aimed at *Newsweek*'s own coverage of the WTO protests, which included a colour photograph, covering nearly two pages of the magazine, depicting a number of individuals exiting a Starbucks coffee shop through smashed windows, and clambering over the upturned stools and tables. This was a photograph carried by a great number of mainstream media outlets. In *Newsweek* the photograph was accompanied by a caption which ran: "rampaging protestors attacked any symbol of globalisation – like this Starbucks shop".[86]

The caption served ostensibly to explain and elucidate the meaning of the image, but, of course, it was actually as much about constructing meaning as making any inherent meaning 'clear'. The appeal of the photograph lay both in its dramatic potency, and its seeming usefulness in distilling complex social and political issues. The choice of Starbucks over and above a number of other damaged downtown retail outlets in Seattle suggested something more. Like a number of the other retailers whose premises were damaged, such as FAO Schwartz, Planet Hollywood, Nike Town, Banana Republic or Old Navy, Starbucks was a 'high-end' brand-led retailer whose employment of an international division of labour within its production processes had come under scrutiny by a range of protest and activist groups in the 1990s.[87] Its high visibility and brand kudos, as well as its 'ethical' corporate image, thus made it a potent target. Yet it was Starbucks's cultivation of a close association with Seattle that made the picture of its attack so appealing to a range of national and international media outlets. The company's own desire to combine the symbolic local with the global had been exploited by newspaper and magazines keen to grasp the local and the global in one spectacular image.[88]

Whilst the history of coffee points to the fact that "in some senses globalisation is a very old process", the emergence of gourmet coffee as a new consumption practice was intimately connected to the profound socio-spatial restructuring of America's urban centres over the last twenty years.[89] These processes of restructuring, and the concomitant growth in importance of the city's symbolic economy, have themselves to be situated within the context of a global political economy signified by greatly intensified patterns of flows for capital, commodities, information and images, and the advanced flexibility of labour and markets. The desire to generate notions of 'the local' and the 'place-specific', needs, as Stuart Hall points out, to be situated "within the logic of globalisation". The attempts by gourmet coffee retailers to mobilise images of Seattle on a global scale exemplifies this process at work. However, the complexity of the global connections epitomised by a commodity such as coffee meant that Starbucks' desire to control the connection between gourmet coffee and Seattle – in their words to project "where we came from" became an increasingly difficult one within a period characterised by a shift in the politics of food consumption and closer scrutiny of the relations of commodity production. Ironically then, while gourmet coffee as a niche consumption practice has prospered from mobilising largely invented ties to Seattle through a savvy process of intertwining 'cultural symbols and entrepreneurial capital', the WTO demonstrations saw the more tangible connections to place embedded in the post-colonial power relations of the global coffee trade inscribed into the material landscape of that very city. [90] If Starbucks profited from inviting customers to 'eat the street', then events in Seattle witnessed the streets bite back.

UNIVERSITY COLLEGE WINCHESTER
LIBRARY

CONCLUSION

In the aftermath of the civil unrest that accompanied Seattle's hosting of the WTO Ministerial Conference Meeting, city mayor Paul Schell was quoted assuring the world's media that "that was *not* Seattle", a declaration that sought to dissociate the city from its shocking tableaux.[1] Having regained authority over the streets of Seattle, Schell's words revealed his wish to similarly wrest control of the image his city would project to national and international viewing publics. Yet even as Schell spoke, the mainstream press was in the process of re-evaluating the city. *USA Today* noted the shock of seeing tear gas wafting "over this usually laid-back city",[2] while the *New York Times* stressed the extraordinary scenario of martial law in the "happiest, mellowest city on the planet".[3] In addition, *Newsweek* stated that "to a lot of people in upwardly-mobile Seattle, the meeting had seemed like a great way to showcase their city",[4] while the British *Independent* noted that for a city of "conscientious citizens … the fervour of demonstrations, the violence that broke out on the fringes, and the police response were an alarming and unfamiliar sight", adding that "Seattle may never be the same again."[5]

Within eighteen months of the events of the WTO meeting, the Nasdaq index crash that initiated the global collapse in dot-com equities would further tarnish the lustre of Seattle and its status as a symbol of national success and urban vitality. In actual fact that process could be argued to have begun with the much-publicised layoffs at the Boeing Airplane Company in December 1998. As the company announced plans to cut production of the 747 aeroplane to one a month by the end of the decade and make 48,000 of its workforce redundant, the London *Times* reported that Boeing workers were being handed yellow leaflets with information on "what to do when you're laid off", commenting that "it was hardly a scene from the booming high-tech America that Seattle is meant to epitomise".[6] In fact the *Times* made much news of the Boeing layoffs, even including a picture of the cast from *Frasier* against the backdrop of the city skyline, accompanied by the caption "From cool to chill: the urban chic of Seattle portrayed in *Frasier* is giving way to the icy blast of recession."[7]

Taking a longer view, it is possible to argue that Seattle's trajectory in the 1990s was simply in keeping with the city's historic cycles of boom and bust, with the end of the decade heralding a predictable fall from grace. Yet this is also to miss much of the specificity of what Seattle signified in the 1990s, particularly in the first half of the decade, which witnessed the city's most intensive phase of coverage and exposure. In retrospect is it clear to see how Seattle's representational currency at this time was allied to a specific cultural moment, and mapped the interplay of a range of anxieties, hopes and desires pertaining to the conditions in America's large urban centres. Apparent in reflecting back upon the dates of the selected texts examined in this book is how much they coalesce around the period from the late 1980s to the mid-1990s. This is not to say that there are not important or popular texts examined here that stray beyond these temporal parameters – *Frasier*, for example, began during that period but endured way beyond. However, as Robert Beauregard notes, the period between the late 1980s and mid-1990s was also one that witnessed "a new economic recession placing cities such as New York and Philadelphia on the edge of bankruptcy and the riots in Los Angeles bringing race urgently to the forefront of concerns over urban decline", a time when, as Liam Kennedy points out, Los Angeles and New York "began to appear in cultural representations as emblematic sites of strain and fracture in the symbolic order of the national culture".[8] That this period corresponded with the generation of many of the texts of Seattle examined here is informative, and reinforces a sense of why the city's appeal was particularly great at this time.

Neither Beauregard nor Kennedy would suggest that urban crisis was the only thematic extant in the cultural representations of cities such as Los Angeles and New York at this time. Similarly, I am not claiming that the only way in which Seattle was represented during this period was in order to valorise it as an alternative and more appealing urban centre. The surfeit of narratives and images about any large city make the idea of comprehensive coverage or smooth consensus absurd. However, the instancing of noteworthy trends and dispositions within the city's signifying repertoire is viable, and what is clear is that a number of the most popular and significant representations of Seattle can be seen to be particularly responsive to crucial issues pertaining to contemporary urbanism.

Looking at the range of selected texts and cultural commodities examined in this book it is also apparent how many of them were tailored towards so-called 'high-end' audiences and consumers: 'quality television' programmes such as *Frasier*, *Twin Peaks*, *Northern Exposure* and *Millennium* aimed at upscale 'niche' audiences; quality 'niche' Seattle-signified commodities such as outdoor apparel, northwest 'lifestyle' products and gourmet coffee; magazines aimed at upscale readerships such as *Time*, *Newsweek*, *The Christian Science Monitor* and *Vogue*; and 'A-class star vehicle movies' such as *Disclosure*, *Sleepless in Seattle* and *Fight Club*. In an important sense this book has served to historicise and contextualise the appeal of Seattle to what could, in Barbara Ehrenreich's terms, be called the 'new middle class' or the 'new professional-managerial class'. Clearly, it would take detailed ethnographic research beyond the scope of this study to determine whether it was indeed predominantly this cohort who engaged with these texts and commodities, and it would require further research to ascertain the nature of those engagements. I am certainly not claiming that Seattle only appealed to this delimited constituency, nor suggesting that other cities have not been the subject of representations aimed at this demographic. However, what is evident is that a proliferation of popular texts and lucrative new consumption practices seeking to foreground Seattle during this period, and aimed at the 'new middle class', can be situated meaningfully against the backdrop of renewed concerns over urban decline. What this book has sought to do is to elucidate and critique that appeal, and to position it within the context of the profound socio-economic restructuring of America's large urban centres. Whilst attentive to the ways in which selected texts represented important instances of contradiction or subversion, the focus has therefore been in interrogating, historicising and contextualising the prevailing prejudices and predispositions inherent in the city's appeal to a particular demographic.

Although a large part of this study has been taken up with delineating the ways in which Seattle came to connote a more desirable vision of American urbanism, and thus functioned ostensibly as a city that had 'avoided' urban decline and its myriad indicators, Seattle's signifying function should most properly be understood as symptomatic of a 'crisis of urbanity' rather than its chimerical solution. Writing about the broader dynamics of urban restructuring in the late 1980s and early 1990s, Kevin Robins refers to the crucial role performed by "that obscure object of desire that is called urbanity".[9] Robins cites developments such as the revalorising of the 'regional and locality', obsessions with 'liveability', the 'soft technological urbanism' promised by computer and communications technology, and the emphasis on new metropolitan consumption practices as symptoms of urbanity 'in crisis' – in the sense that they failed to address adequately "the real complexities and contradictions" thrown up by "the increasing fragmentation and segmentation of urban life".[10] As the themes of this study suggest, during the period in question such concerns found their organising site and symbolic repository in the shape of Seattle. Whilst such themes might have functioned discursively to signify the city's distinctiveness, in fact they evinced "the dynamic relationship between the universal and the particular".[11] Nowhere was this more evocatively in evidence than in the search by a number of national and international periodicals in the mid-1990s to find "the next Seattle". The phrase underlined something of the paradox of place identity within what Robins terms the 'global-local nexus' – with the local being understood as "a fluid and relational space, constituted only in and through its relation to the global".[12] At the same time, it reflected the preference on the part of some sections of the American media, to quote Sam in *Sleepless in Seattle*, to find "a new city", rather than confront the difficulties posed by the old.

NOTES

INTRODUCTION

1 'The Storm: Part II, *E.R*, Channel 4, 19 May 1999.
2 Roger Sale, *Seattle, Past to Present* (Seattle: University of Washington Press, 1976), 94.
3 Sale 1976: 94.
4 Ibid.
5 For more information, see Carl Abbott, *The Metropolitan Frontier* (Tucson: University of Arizona Press, 1993).
6 John M. Findlay, 'The Seattle World's Fair of 1962: Downtown and Suburbs in the Space Age', in *Magic Lands: Western Cityscapes and American Culture Since 1940* (Berkeley: University of California Press, 1992), 215, 218.
7 Sale 1976: 232.
8 *Stage Door*, dir. Gregory La Cava, perf. Katherine Hepburn, Ginger Rogers and Lucille Ball, RKO, 1937.
9 Clark Humphrey, *Loser: The Real Seattle Music Story* (Portland: Feral House, 1995), 2. In addition, Carlos Schwantes maintains that not until "the 1950s and 1960s would the Pacific Northwest cease being viewed by most outsiders as an artistic backwater content with a largely derivative cultural life and begin to develop its own distinctively regional style of painting and literature." Carlos Schwantes, *The Pacific Northwest* (Lincoln: University of Nebraska Press, 1996), 307.
10 J. D. Salinger, *The Catcher in the Rye* (London: Penguin, 1958), 79.
11 For more on this subject see Randy Hodgins and Steve McLellan, *Seattle on Film* (Olympia: Punchline Productions, 1995).
12 Conversation between Jim McLean (Guy Boyd) and Sam (Gregg Henry) in *Body Double*, dir.

Brian De Palma, Columbia Pictures, 1984.

13 'Swimming to Seattle', *Newsweek*, 20 May 1996: front cover.

14 For two contrasting accounts of the troubled identities and possible solutions for America's urban malaise before the period currently under analysis, see Jane Jacobs' seminal *The Death and Life of Great American Cities* (London: Jonathan Cape, 1962) and Louis Mumford, *The City in History: Its Origins, Its Transformations, and Its Prospects* (New York: Harcourt, Brace and World, 1961).

15 James Donald, 'Metropolis: The City as Text', in Robert Bocock, Robert Thompson and Kenneth Thompson (eds), *Social and Cultural Forms of Modernity* (Cambridge: Polity Press, 1992), 452.

16 For more information on the transforming nature of the American city during this period, see Michael R. Dear, *The Postmodern Urban Condition* (Oxford: Blackwell, 2000); Sharon Zukin, *Landscapes of Power: From Detroit to Disneyland* (Berkeley: University of California Press, 1992); Sharon Zukin, *The Culture of Cities*, (Oxford: Blackwell, 1995); M. Christine Boyer, *The City of Collective Memory* (Cambridge MA: MIT Press, 1994); or Robert A. Beauregard, *Voices of Decline* (Cambridge MA: Blackwell, 1993). For work on the processes of gentrification, see Neil Smith, *The New Urban Frontier* (London: Routledge, 1996); Neil Smith and Peter Williams (eds), *Gentrification of the City* (Winchester MA: Allen and Unwin, 1986).

17 Mike Davis, *City of Quartz* (London: Vintage, 1990), 224.

18 Liam Kennedy, *Race and Urban Space in Contemporary Culture* (Edinburgh: Edinburgh Univerysity Press, 2000), 17.

19 Mike Davis, *Ecology of Fear* (London: Picador, 1999), 276, 282.

20 Much research has been undertaken on the relationship between the city and the cultural work of representation in recent times. For instance, the work of Michel de Certeau on the metaphoric, mythical, labyrinthine, experiential city is exemplary in this regard, as is Donna Mazzoleni's writing on the city as a site for imaginary, symbolic desires. (See Michel de Certeau, 'Walking the Streets', in *The Practice of Everyday Life* (Berkeley: University of California Press, 1988); Donna Mazzoleni, 'The City and the Imaginary', in Erica Carter, James Donald and Judith Squires (eds), *Space and Place: Theories of Identity and Location* (London: Lawrence and Wishart, 1993), 285–302. See also M. Christine Boyer, 'The City as Illusion: New York's public places', in P. Knox (ed.), *The Restless Urban Landscape* (New Jersey: Prentice Hall, 1993), 111–26; M.Christine Boyer, *Dreaming the Rational City* (London: MIT Press, 1986). There has also been a 'rediscovery' of earlier studies of late nineteenth- and early twentieth-century cities, in particular to translations of Walter Benjamin's work on the rebuilding of Paris under Baron Haussmann, the vision of the same city offered by the poet Charles Baudelaire, and a revival of interest in the writings of Georg Simmel (see, for example, Marshall Berman, *All That Is Solid Melts Into Air* (London: Verso, 1983). Berman draws heavily upon both Baudelaire and Haussmann in order to construct a compelling account of the experience of modernity, and to connect what he calls "the dynamic and dialectical modernism of the nineteenth century" with the contemporary period (36)). For a contemporary application of Georg Simmel's writing, see James Donald, 'The City, The Cinema: Modern Spaces', in C. Jenks (ed.), *Visual Culture* (London: Routledge, 1995). Much of this work has sought to address the lacunae in macro-level urban studies, particularly with regard to issues of race, sexuality and gender – Elizabeth Wilson's *The Sphinx in the City*, for example, takes a keen eye to what she sees as the coded gendering of city culture and urban space (Elizabeth Wilson, *The*

Sphinx in the City: Urban Life, the Control of Disorder, and Women (London: Virago, 1991)).
Indeed, the new directions provided by the injection of art or literary criticism, and the study of
written, painted, filmed and photographed representations of the city, has led to new theoretical
and conceptual synergies and dialogues around the burgeoning cross-disciplinary field
of Visual Culture. See, for example, T. J. Clark, *The Painting of Modern Life: Paris in the Art of
Manet and His Followers* (London: Thames and Hudson, 1984); Griselda Pollock, *Vision and
Difference: Femininity, Feminism and the Histories of Art* (London: Routledge, 1988); Janet Wolff,
'The invisible *flaneuse*: women and the literature of modernity', *Theory, Culture and Society*,
2: 3; John Tagg, *The Burden of Representation: Essays on Photographs and Histories* (Minnesota:
University of Minnesota Press, 1993).

21   Beauregard 1993: xi
22   Beauregard 1993: 13.
23   Beauregard 1993: 220.
24   Beauregard 1993: 231. See, for example, George J. Church, 'Urban Growing Pains: Denver
      decides to take off, but booming Seattle hunkers down', *Time*, 29 May 1989: 33; 'Town
      Planning: Where it Works', *The Economist*, 1 September 1990: 36, 39; 'Cities that Satisfy,
      American Demographics, September 1995: 18–20; Anne Fisher, 'The Best Cities for Busi-
      ness', *Fortune*, 20 December 1999: 102–6. For more in depth analysis see Mark Gottdiener,
      Claudia C. Collins and David R. Dickens, *Las Vegas: The Social Production of an All-American
      City* (London: Blackwell, 2000); Charles Rutheiser, *Imagineering Atlanta* (New York: Verso,
      1996).
25   'Citistate resurgent', *The Economist*, 13 November 1993: 71; Alan Artibise, Anne Vernez
      Moudon and Ethan Seltzer, 'Cascadia: An Emerging Regional Model', in Robert Geddes
      (ed.), *Cities in Our Future* (Washington DC: Island Press, 1997): 149–74.
26   For an informative analysis of the corresponding, and often intertwined histories of Seattle
      and Vancouver, see Norbert MacDonald, *Distant Neighbors* (Lincoln: University of Nebraska
      Press, 1987).
27   James Lyons, '"The Manson Family Used to Vacation up Here": Northwestploitation, Micro-
      cinema and the Battle of Seattle'. Conference paper delivered at *Defining Cult Movies: The
      Cultural Politics of Oppositional Taste*, Broadway Cinema, Nottingham, UK, 17 November,
      2000.
28   Implicit in the approach taken by this study is a desire to chart what could be characterised
      as the dominant orientation of these key themes as they pertain to Seattle. The notion of
      'dominance' within discursive formations is a complex idea, and requires some explanation.
      Inherent in the study's notion of dominance within discourse is the concept of hegemony,
      developed by the Marxist theorist and political activist Antonio Gramsci in the 1930s, and
      later taken up within the field of cultural studies. The main contribution of Gramsci's
      concept of hegemony, as it has been applied within cultural studies, is to provide a more
      subtle and nuanced explanation for how dominant social groups achieve and maintain their
      dominance. Unlike more orthodox Marxist formulations of ideology, Gramsci's formulation
      does not depend on a notion of coercion or manipulation, but rather seeks to show how
      "consent is actively sought for those ways of making sense of the world which 'happen' to
      fit with the interests" of the hegemonic group. Moreover, this struggle for consent is won
      (and must be continually re-won) at the level of popular culture – it is the cultural work
      of representation that provides "the area of negotiation" (John Hartley, 'Hegemony', in

Tim O'Sullivan, John Hartley, Danny Saunders, Martin Montgomery and John Fiske, *Key Concepts in Communication and Cultural Studies* (London: Routledge, 1997), 133. For an examination of Gramsci's theory of hegemony, see Quintin Hoare and Geoffrey Nowell-Smith (eds), *Selections from the Prison Notebooks of Antonio Gramsci* (London: Lawrence and Wishart, 1973). For an account of the return to the work of Gramsci within cultural studies, see Tony Bennett, Colin Mercer and Janet Woollacott (eds), *Popular Culture and Social Relations* (Milton Keynes: Open University Press, 1986).

29 Ann Japenga, 'On a Northwest Course', *Los Angeles Times*, 24 December 1992: E8.

CHAPTER 1

1 Details from www.cityofseattle.net/tda/industry/software_it.htm.

2 Sharon Zukin, *Landscapes of Power: From Detroit to Disneyland* (Berkeley: University of California Press, 1992), 4.

3 Carlos Arnaldo Schwantes, *The Pacific Northwest* (Lincoln: University of Nebraska Press, 1996), 507.

4 Bradley Meacham and David Woodfil, 'So Long Seattle: More People Are Moving Away', *Seattle Times*, 4 January 2003.

5 John M. Findlay, 'The Seattle World's Fair of 1962: Downtown and Suburbs in the Space Age', in *Magic Lands: Western Cityscapes and American Culture since 1940* (Berkeley: University of California Press, 1992), 214.

6 Findlay 1992: 256.

7 Mary Bruno, 'Seattle Under Siege: The last best city in America fights for its soul', *Lear's*, July 1991: 49.

8 Bruno 1991: 91.

9 Ibid.

10 For a useful discussion of this recession, and its impact, see Beauregard 1993.

11 Bruno 1991: 51.

12 Mike Davis, *City of Quartz* (London: Vintage, 1990), 224.

13 Clearly, given the contrast with Los Angeles and New York, there was also a racialised dimension to this vision of Seattle on the edge, and this is something picked up on in chapter three.

14 *Newsweek*, 26 May 1996: 57.

15 Jerry Adler, 'Seattle Reigns', *Newsweek*, 26 May 1996: 48.

16 Adler 1996: 49.

17 Adler 1996: 50.

18 Donald Lyons and Scott Salmon, 'World Cities, multinational corporations, and urban hierarchy: the case of the United States', in Paul L. Knox and Peter J. Taylor (eds), *World Cities in a World-System* (Cambridge: Cambridge University Press, 1995), 99.

19 Davis 1999: 360. Indeed, such past imaginings can be seen by examining *It Happened at the World's Fair* (1963), a motion picture filmed around the staging of the Seattle World's Fair in 1963. The movie was a vehicle for Elvis Presley, who played Mike Edwards (appositely for the home of Boeing, a pilot) who, along with his co-pilot, hitchhikes a ride to Seattle to look for work that would earn them the money to buy back their impounded plane, and happen to 'stumble upon' the World's Fair. The movie's minimal plot allows ample opportunities for

Elvis to sing songs and to court his obligatory love interest, in this case, nurse Diane Warren (Joan O'Brien), who is working at the Fair. It also enables the World's Fair to take a starring role, in the sense that the couple's courtship is conducted primarily against the backdrop of the exposition: multiple rides on the monorail, dinner in the Space Needle restaurant, and scenes amongst the Space Arches at the US Science Pavilion. The movie suggests that Seattle and the World's Fair are synonymous – the city does not exist but as a landscape of the future and the architectural icons of future technology inscribe themselves literally onto the fabric of the city. Referring to nurse Warren's stated desire to join NASA's "space medicine program", Mike (Presley) professes his admiration and affection for her as someone who "look[s] ahead to the future". Indeed, so strong is the intoxicating mixture of Century 21 ethos and nursing charm that by the movie's dénouement, Mike has also decided to put his own pilot skills to more forward-looking endeavours and signs up "for something in the space program" at the NASA Century 21 stall. *It Happened at the World's Fair*, filmed at the end of the economic honeymoon of the baby-boom era and brimming with optimistic civic boosterism, implied that the future (and by association the future of Seattle) was benign, romantic, wondrous and deeply middle-class.

20 Dean Paton, 'Seattle's rise as the capital of the New Economy', *Christian Science Monitor*, 30 November 1999: 1–4.

21 Paton 1999: 4.

22 Ibid.

23 Ibid.

24 Ibid.

25 For example, *Lear's*, which circulated between 1988 and 1994, was aimed at a readership in the over-40 age bracket, while the *Christian Science Monitor* describes its market to advertisers as a mature, affluent and well-educated demographic, with a median age of 59. *Newsweek*, an internationally distributed news periodical which had a domestic circulation of around 19 million per week during the 1990s, uses its website to inform potential advertisers about its educated, affluent and predominantly middle-aged readership. Details on *Lear's* from 'A Maturing Woman Unleashed', *Time*, 15 May 1989; details on the *Christian Science Monitor* from www;csmonitor.com/aboutus/advprint demog.html, first accessed 4 August 2001; details on *Newsweek* from the *Newsweek* website: www.newsweekmediakit.com/us/reader_national.html, first accessed 1 June 2001.

26 Peter Williams and Neil Smith, 'From "Renaissance" to "Restructuring"', in Neil Smith and Peter Williams (eds), *Gentrification of the City* (London: Unwin Hyman, 1988), 208.

27 Beauregard 1993: xi.

28 *Jurassic Park* was, at the time of its release, the biggest-grossing box-ffice movie in history, making over $912 million worldwide. *Rising Sun* (1993) made $65 million at the domestic box office, *Disclosure* (1994) made $83 million, whilst the critically derided *Congo* (1994) still managed to make $80 million in domestic sales.

29 Gregory Jaynes, 'Meet Mister Wizard', *Time*, 25 September 1995: 34–7.

30 For example, his appearance on the front cover of *Time* magazine in September 1995, timed to coincide with the publication of *The Lost World*, his sequel to the 1990 blockbuster *Jurassic Park*, served to reinforce the sense of Crichton's status as an important figure within contemporary culture.

31 For a discussion of the restoration of Pioneer Square, see Roger Sale, *Seattle, Past to Present*

(Seattle: University of Washington Press, 1994).

32 Michel Crichton, *Disclosure* (London: Arrow, 1994), 20.

33 Constance McLaughlin Green, 'Seattle, City of the Northwest', *American Cities in the Growth of the Nation* (London: Athlone Press, 1957), 168.

34 David M. Wrobel, *The End of American Exceptionalism: Frontier Anxiety from the Old West to the New Deal* (Lawrence: University Press of Kansas, 1993), 145.

35 Robert X. Cringely, *Accidental Empires* (London: Penguin Books, 1996), 15.

36 Cringely 1996: 16.

37 Ibid.

38 Richard Slotkin, *Gunfighter Nation* (New York: Atheneum, 1992), 4.

39 Slotkin 1992: 4.

40 Slotkin 1992: 65.

41 Neil Smith, *The New Urban Frontier* (London: Routledge, 1996), xiv.

42 Ibid.

43 Ibid.

44 Ibid.

45 If, at the end of the twentieth century, such recourse to the terminology of frontiers and pioneers seems, at the very least, a little indolent conceptually then perhaps it is. As Patricia Nelson Limerick points out, "the scholarly understanding formed in the late nineteenth century still governs most of the public rhetorical uses of the word 'frontier'". Patricia Nelson Limerick, 'The Adventures of the Frontier in the Twentieth Century', in Richard White and Patricia Nelson Limerick, *The Frontier in American Culture* (Berkeley and Los Angeles: University of California Press, 1994), 95. The frontier as metaphor in the Cold War was given its force as an analogy in John F. Kennedy's Democratic presidential nomination speech in July 1960, which made powerful reference to the New Frontier of the 1960s. Yet it may also be particularly apposite, at least as it is applied to Seattle. As John Findlay argues in *Magic Lands*, critics of Western cities have often "clung inflexibly to an increasingly obsolete idea of city life, rooted in the urban experience of the Northeast, in which strong central cities with vital downtowns dominated metropolitan areas ... an Eastern model, based on nineteenth-century technologies" (John M. Findlay, *Magic Lands: Western Cityscapes and American Culture since 1940* (Berkeley: University of California Press), 7). In this way, Seattle's representation as a "city on the frontier" represents, therefore, something of a journey "back to the future", utilising, as it does, not only nineteenth-century terminology but also a nineteenth-century concept of city form.

46 Jaynes 1995: 34.

47 In *Accidental Empires* Robert X. Cringley recounts the battles between IBM, seen as "the establishment" – the inherently conservative corporate America, complete with a conservative dress code – and the Seattle software company Microsoft in the late 1980s. He characterises the relationship between Microsoft and IBM in the 1980s as a culture clash, noting that "IBMers were buttoned-up organisation men, Microsofties were obsessive hackers" (68). The severing of the relationship between the companies in 1989/1990, resulted from the disagreements between them over the future over OS/2, an operating system Microsoft were developing for IBM and intended to replace MS-DOS. The result was Microsoft going it alone, transforming the industry, and reaping huge financial rewards from the Windows graphical user interface it had been developing over the same period. The result was also that

the "faster, more flexible business culture" of Microsoft became the template for future success in the software industry. IBM, famous for moving at a "glacial pace", and requiring years to make any major decision became "an also-ran in the PC business" (76). IBM, as the corporate behemoth that failed to sustain itself on the cutting edge, also, in effect, became a blueprint for how *not* to succeed in the PC business, and served as a warning to other companies in its guise as the "ghost of corporations past".

48   *Disclosure*, dir. Barry Levinson, perf. Demi Moore and Michael Douglas, Warner Brothers, 1994.

49   Geoff Mulgan, 'High Tech and High Angst', in Sarah Dunant and Roy Porter (eds), *The Age of Anxiety* (London: Virago, 1997), 2.

50   Mulgan 1997: 3.

51   Barbara and John Ehrenreich, 'The professional-managerial class', *Radical America*, Part 1, 11 (March–April 1977): 7–31; Part 2, 11 (May–June): 7–22.

52   Barbara Ehrenreich, *Fear of Falling* (New York: Pantheon Books, 1989), 15.

53   Ibid.

54   Ibid.

55   Sharon Zukin, *Landscapes of Power: From Detroit to Disneyland* (Berkeley: University of California Press, 1992), 3.

56   Kennedy 2000: 42.

57   Smith and Williams 1986: 11.

58   Molly Dee Anderson, 'North of the City', in John Wilcock (ed.), *Insight Guides: Seattle* (Singapore: APA Publications, 1997), 197.

59   'Technology Corridor', www.sscchamber.org./Area/TC.html.

60   Schwantes 1996: 435.

61   Bellevue played host to a three-day conference on the 'Emerging City' in September 1991, sponsored by the city of Bellevue and the University of Washington. Joel Garreau led the panel discussion at the conference.

62   Joel Garreau, *Edge City* (New York: Doubleday, 1991).

63   Garreau 1991: 4.

64   Garreau 1991: xxii, 4. It should be added that Garreau's conception of the 'edge city' is far from unproblematic, not least in its unbridled optimism and enthusiasm for such ex-urban agglomerations in the face of some quite brazen class bias. For example, Garreau argues for the new locale's improvement over the "old suburbia-downtown arrangement" because "it moves everything closer to the homes of the middle class" (9). Where that would leave those who are left behind in the downtown remains largely unexamined.

65   For more information on this, see the passage on *Generation X* in chapter five.

66   See for example, *Shampoo Planet* (1992), which focuses on teenagers of the 1990s; *Life After God* (1994), a short story collection emphasising twentysomething existential crises; and *Girlfriend in a Coma* (1997), a bleak novel of high-school sweethearts.

67   Douglas Coupland, *Microserfs* (London: Flamingo, 1995), 3.

68   It is also important to note that Coupland's characters come to understand that Microsoft's maturation as a company means they will never become stock-holding millionaires, or Cyberlords, and will forever (until downsized) remain Microserfs. What is fascinating about *Microserfs* is the way that its image of living on the 'cutting edge' of technology portrays paradoxically a deeply backwards culture. Cyberlords and Microserfs are words that combine

the futuristic and the feudal: the industry may be on the frontier of the new world, and, in Joel Garreau's terms, the frontier of "how cities are being created", but the corporate culture evokes, anachronistically, the old world, of deference, demagoguery and kingdoms.

69 The irony is that Dan and friends leave Microsoft and Seattle for Palo Alto, California, the very template of the shape and form of Seattle's future development that the edge city represents, and which many commentators would wish it to avoid. For example, in an article in the 7 July 1997 edition of *Fortune* magazine, Mark D. Fefer posed the question 'Is Seattle the next Silicon Valley?', *Fortune*, 7 July 1997, www.pathfinder.com/@@Btj*4AUAotoeCma9/1997/970707/sea.html). The article goes on to suggest that "Each high-tech region has its unique appeal, but the chemistry is especially potent in Seattle. Combine the grand ambitions and colossal wealth of people like [Paul] Allen and Bill Gates with a ferment of new ideas and entrepreneurial energy, then add in the towering strength of Microsoft and a quality of life that is envied by much of the country, and you have the fixings, some observers believe, for a new Silicon Valley." Fefer then suggests that in spite of "the chemistry … there's little evidence that the locals have any desire to see Seattle become the next high-tech mecca". Local fears over housing prices and traffic congestion reaching "California proportions" lead Fefer to conclude that if "Seattle is destined to become the next Silicon Valley, it will probably be in spite of itself". Here, again then, is the 'catch 22' of urban development engendered by the influence of advanced tech industries upon the area. On the one hand, advanced technology enables Seattle to be a site of pioneering endeavour, to avoid the perils of the 'older and stodgier' large industrial cities of the northeast. On the other hand, being 'on the edge' is understood to involve the peril of falling into another pattern of urban decline, namely, the sprawl and congestion typified by the example of Silicon Valley.

70 Coupland 1995: 4.

71 Coupland 1995: 7.

72 Coupland 1995: 23.

73 Coupland 1995: 15.

74 Kevin Robins, 'Prisoners of the City: Whatever Could a Postmodern City Be?', in Erica Carter, James Donald and Judith Squires (eds), *Space and Place: Theories of Identity and Location* (London: Lawrence and Wishart, 1993), 326.

CHAPTER 2

1 David Bell and Gill Valentine, *Consuming Geographies* (London: Routledge, 1997), 148.

2 Jonathan Raban, quoted in 'Seattle Reigns', *Newsweek*, 26 May 1996, and in *Cities of the Future* (BBC, 1996). For examples of Raban's writing on the Pacific Northwest, see the aforementioned *Hunting Mr Heartbreak* (London: Collins Harvill, 1990); *Passage to Juneau* (London: Picador, 2000); or 'Battleground of the Eye', *The Atlantic Monthly*, March 2001.

3 Raymond Williams, *The Country and the City* (London: Hogarth Press, 1973), 289.

4 Beauregard 1993: 14.

5 Beauregard 1993: 15.

6 See, for example, James L. Machor, *Pastoral Cities: Urban Ideals and the Symbolic Landscape of America* (Madison: University of Wisconsin Press, 1987); John F. Kasson, *Civilizing the Machine: Technology and Republican Values in America, 1776–1900* (New York, 1976); and Leo Marx, *The Machine in the Garden: Technology and the Pastoral Ideal in America* (Oxford:

Oxford University Press, 1964).

7   Alexander Wilson traces the development of such initiatives in *The Culture of Nature* (Cambridge MA: Blackwell, 1992). For example, he refers back to the Volksgarten, Europe's first established public parks, which started as private reserves before gradually allowing public access by the end of the eighteenth century. Wilson also makes reference to the translation of the English landscape park, underpinned by a pastoral view of nature "as a kindly mother" and refuge, to North America from the European conquest onwards, and cites Frederick Law Olmstead's Central Park in New York City as one of the most striking examples. Interestingly, in 1903 the Seattle Park Board hired the sons of Frederick Law Olmstead to design the city's park system. The unbuilt centrepiece of the Olmsteads' vision was a similarly large open space in the heart of the city, which underpinned the Olmstead philosophy "that a park was essential to the health of the community" (Lynne B. Iglitzin, 'The Seattle Commons: A Case Study in the Politics and Planning of an Urban Village', *Policy Studies Journal*, 23: 4, 1995: 620–35).

8   Davis 1999: 64.

9   Davis 1999: 72.

10   Davis 1999: 318.

11   Ibid.

12   One thing that is clear from the proliferation of narratives of 'nature' that have accompanied and shadowed the rise of environmental movements in the last thirty years is the connotative versatility of the term. Indeed Alexander Wilson repeats Raymond Williams' adage that "nature [is] the most complex word in the language" (1992: 12). At a theoretical level, critical inquiries into the discursive construction and identity of 'nature' and 'the natural' by feminism, post-structuralism and deconstruction have also made it "impossible to maintain previous understandings of 'nature' as an innocent given" (George Robertson, 'Introduction: look who's talking', in *Futurenatural* (London: Routledge, 1996), 2). As Neil Smith suggests, commenting on the recent 'discovery of nature' by cultural studies, nature has always been *in* literary and cultural texts, yet for the main has been "rendered a backdrop, a mood setter, at best a refractory image of, or rather simplistic metaphor for, specific human emotions and dramas that inscribe the text" (Neil Smith, 'The Production of Nature', in George Robertson (ed.), *Futurenatural*, 42). Yet if we are now seeing the trees in the forest, it has become more difficult to put a finger precisely on the identity of *what* we are seeing. 'Nature' has been revealed as a fabulously nebulous and slippery term – whilst it has a referent 'out there', our understandings of it are most assuredly shaped by discourse – as Wilson points out, "nature is part of culture" (ibid.).

13   John Spengler and Tim Ford, 'From the Environmentally Challenged to the Ecological City', in Robert Geddes (ed.), *Cities in Our Future* (Washington DC: Island Press, 1997), 33–4.

14   See note 13, above. A more popular instance of such sentiment is Anne Matthew, *Wild Nights: Nature Returns to the City* (New York: North Point Press, 2001). Matthews explores the ways in which wild animal life – what she calls "a resurgent natural world" – has recently reasserted itself within the nation's most urbanised zones.

15   Wilson 1992: 12. As Robertson explains, "There are real threats to whatever we conceive the 'natural' to be; the air, the land, the oceans, our bodies. The fragility of the concept 'nature' and the instability of its referent are strikingly demonstrated in the muddled vocabularies of environmental politics and in our struggles to come to terms both with the implications

of new genetic and reproductive technologies and with the psychic consequences of the loss of 'nature' as a foundational concept, a ground of being, a stable otherness to the human condition" (1996: 2).

16 For example, Jonathan Raban's assertion that "people come to Seattle because of the mountains, because of the lakes, because of the Puget Sound … they wanna hike, they wanna bike, they wanna sail". Such people, Raban posits, regard Seattle's main function as "a point of access to a gigantic natural park" (Raban, quoted in *Cities of the Future: Seattle*, Exec Prod. Alex Graham, BBC Television, 1996).

17 Sophie Davies, 'Speechless in Seattle', *She*, September 1999: 196; JoAnn Roe, 'Seattle: The Emerald City', *American West*, April 1989: 42; Jerry Adler, 'Seattle Reigns', *Newsweek*, 20 May 1996: 53.

18 'Wired in the Woods', *Economist*, 31 July 1999: 44–5.

19 Carlos Arnaldo Schwantes, *The Pacific Northwest* (Lincoln: University of Nebraska Press, 1996) 476–7. As Schwantes points out, whilst most inhabitants and commentators would agree on the 'core' states of Washington and Oregon, "some would include western Montana and even Northern California and British Columbia within the region" (2). For my purposes, the fact that Seattle remains within virtually all definitions will suffice.

20 Schwantes 1996: 477.

21 Ibid.

22 John M. Findlay, 'A Fishy Proposition: Regional Identity in the Pacific Northwest', in Michael C. Steiner and David M. Wrobel (eds), *Many Wests* (Lawrence KA: University Press of Kansas, 1997), 49.

23 Findlay 1997: 53.

24 Ibid. Findlay notes that "in the first decade of the twentieth century, the Great Northern and Northern Pacific lines … commissioned Tacoma artist Abby Williams Hill to paint mountain scenes. They sent her to work in selected areas, incorporated the resulting paintings into their publicity materials, and displayed her artwork at world fairs. Such efforts increasingly equated the region with its scenery."

25 Findlay 1997: 55.

26 Findlay 1997: 53.

27 Findlay 1997: 54.

28 Schwantes cites the Washington legislature's creation of the Department of Ecology in 1970 as indicative of changing attitudes (1996: 480).

29 Schwantes 1996: 479.

30 Eric Lucas, 'Timber Town', in John Wilcock (ed.), *Insight Guides: Seattle* (Singapore: APA Publications, 1997), 94.

31 Lucas 1997: 92.

32 Ibid.

33 Schwantes 1996: 521. For example, in July 1999, the *Economist* profiled Issaquah, a town of 10,000 people to the Eastside of Lake Washington in the Seattle Metropolitan area, as part of a series on urban growth entitled 'Wired in the Woods'. The article discussed Issaquah's identity as a site for 'green-collar' jobs that create "no smokestacks, [have] no pipelines dumping toxic waste into streams, and [suck] up only as much energy as fluorescent light bulbs and PCs consume". It stated that such 'green collar' employment recruited individuals who "are almost universally green-minded … eager to see forests and streams preserved so they can

hike, ski, kayak and mountain bike". The article went on to note that "unfortunately, those same people also demand five-bedroom houses, cars … (two-ton-plus, four-wheel-drive), and plenty of places to shop". The result was that the "wooded hills that surround Issaquah [are overwhelmed with] subdivisions jammed with $500,000 homes" ('Wired in the Woods', *Economist*, 31 July 1999: 44–5).

34 Schwantes 1996: 521. As Wilson points out, 'nature tourism' involves not simply reorganising the 'uses' of nature, but represents a radical shift in "our perceptions of nature and our place in it as humans", as the land is re-defined "in terms of leisure". Although participation in nature tourism saw its first major spurt of growth in the post-World War Two period, reflecting growing affluence and mobility, the vision of a recreational nature to be "revered as a source of aesthetic pleasure" finds its origins in the upper middle class of the late nineteenth century, when "wealthy city-dwellers were taking curative holidays at Rocky Mountain spas and seaside resorts". Wilson writes that such "nature tourism … is simply the temporary migration of people to what they understand to be a different and usually more 'pure' environment. It's going out to nature for its own sake, and it's all the ways we talk about that experience. The modern history of nature tourism is a history of altered land-forms and changed ideas and experiences of the non-human. Broadly speaking, it involves a shift from a pastoral approach to nature to a consumer approach" (1992: 24).

35 '21 terrific outdoor adventures', *Seattle Magazine*, July/August 1999, front cover; *Seattle Magazine*, 'Editorial Guidelines: target reader', 4 January 2000, www.seattleinsider.com/ partners/seattlemag/editguide.html.

36 Laura Slavik, 'So close, yet it feels so far', *Seattle Magazine*, July/August 1999: 27.

37 This is a conceptualisation of the city's identity that many profiles of Seattle were happy to perpetuate. For example, the *Newsweek* cover article on Seattle from May 1996 carried a quote from the *Seattle Weekly*'s Katherine Koberg, who stated that "I've met a lot of people who are doing impressive things in business, the arts, in academia, but they all say the same thing: 'Oh, I came here because I love the mountains.'"

38 Schwantes 1996: 503. In some senses, the answer to his question depends on who is doing the asking. Take, for example, a cartoon from the Seattle alternative newspaper the *Stranger* on 12 July 1997. Drawn by Dominic Rappello, the cartoon was part of a commentary on the Seattle International Film Festival (SIFF); one of a number of prominent international events that has supposedly served to consolidate the sense of Seattle as a significant nexus of cultural influence. The cartoon depicted local cinephiles quizzing a visiting film director, and thanking her for creating a celluloid Seattle they felt was in tune with their aspirations to sophisticated urbanity. The film director, with plane ticket in hand, had a 'think bubble' drawn above her head, in which the words "get me out of this fishing village" had been written. The cartoon served to mock Seattleites' cosmopolitan pretensions, countering them with a view from 'outside' (where the visiting film director was from was not made clear) that 'saw' Seattle as a parochial maritime locale.

39 Beauregard 1993: xi.

40 Sharon Zukin, *Landscapes of Power: From Detroit to Disneyland* (Berkeley: University of California Press, 1992), 12. See also Arjun Appadurai, 'Disjuncture and Difference in the Global Cultural Economy', *Theory, Culture and Society* (London, Newbury Park and New Delhi: Sage, 1990), 7: 295–310.

41 Kevin Robins, 'Prisoners of the City: Whatever Could a Postmodern City Be?', in Erica

Carter, James Donald and Judith Squires (eds), *Space and Place: Theories of Identity and Location* (London: Lawrence and Wishart, 1993), 304–5.

42 Robins 1993: 304.

43 David M. Wrobel and Michael C. Steiner, 'Many Wests: Discovering a Dynamic Western Regionalism', in David M. Wrobel and Michael C. Steiner (eds), *Many Wests* (Lawrence KA: University Press of Kansas, 1997), 2–3.

44 Wrobel and Steiner 1997: 6–7.

45 Wrobel and Steiner 1997: 8.

46 John M. Findlay, 'A Fishy Propostion: Regional Identity in the Pacific Northwest', in Wrobel and Steiner 1997: 38.

47 Ibid.

48 Findlay 1997: 43.

49 Ibid.

50 As Findlay notes, "Geographers identify two major physiographic provinces … a wetter greener place to the west of the Cascade Range, and a drier, browner place to the east … These coincide with two distinct 'culture areas' … each of which accommodated a set of Indian peoples with certain characteristics not shared on the other side of the Cascades." Additionally, "the more urbanised settlement, timber and fishing economies, and cultural amenities on the ocean side of the mountains … have had no equivalent to the east" (1997: 43).

51 Findlay 1997: 43–4.

52 Findlay 1997: 41. Indeed, a similar claim could be made concerning another recent symbol of the natural Northwest, namely the northern spotted owl. As Carlos Schwantes notes, in June 1990, the owl featured on the front of *Time* magazine, reflecting its status as an endangered species put at risk by "dwindling old growth forests" (1996: 487). The federal Fish and Wildlife service immediately declared the owl a threatened species, precipitating a wave of protests from the logging community. Like the salmon, the owl had gained a new cultural resonance, making it a symbol of an 'essential' 'natural' Northwest that was to be protected, a status which, however worthy, was as mutable as the identity of the region that it served to signify.

53 David Bell and Gill Valentine, *Consuming Geographies* (London: Routledge, 1997), 148.

54 Wrobel and Steiner 1997: 17.

55 Nancy Harmon Jenkins, 'As American as Apple Pie', *New York Times Magazine*, 1 March 1992: 67; Eric Perret, 'A Houseboat in Seattle', *Esquire*, February 1993: 32; Beth Brophy, 'Stressless – and simple – in Seattle', *US News & World Report*, 11 December 1995: 96; Amy Tucker and Stephen Tanzer, 'All of the Best and None of the Worst of Seattle', *Forbes*, 8 May 1995: 146.

56 Wrobel and Steiner 1997: 2.

57 Circulation details from *Roy Morgan International*, "US Paper Readership Estimates", www.roymorgan.com.au/international/mediapapers/2000/mediaMeasurementUS/USNews paperEst_2000.html.

58 Ann Japenga, 'On a Northwest Course', *Los Angeles Times*, 24 December 1992: E1.

59 Ibid.

60 The article even carried a quote from the home store co-ordinator for the Los Angeles Broadway shopping mall, who stated that "without even leaving the mall, Angelenos can

get a Northwest holiday fix. The [mall] is displaying a northern pine tree ornamented for Christmas in tiny lumberjacks, pine cones and a Santa in a plaid grunge-style shirt" (Ibid.).

61 Smith 1996: 35.

62 David Chaney, *Lifestyles* (London: Routledge, 1996), 8–12.

63 Japenga 1992: E2.

64 Ibid.

65 See, for example, David Lipsky, 'Junkie Town', *Rolling Stone*, 30 May 1996: 36–62.

66 Smith 1996: 36.

67 Posner 1997: 10. Posner was quoting Richard Weinstein at the *Conference on Cities in North America* held in New York in 1996. Weinstein went on to point out that "most of the people in the world are poor, and their primary concern is to put bread on the table ... people will choose the right to work over the environment and quality of life as we define it every day of the week in most of the world".

68 Andrew Ross (interviewed by Michael Bennett) 'The Social Claim on Urban Ecology', in Michael Bennett and David W. Teague (eds), *The Nature of Cities: Ecocriticism and Urban Environments* (Tucson: University of Arizona Press, 1999), 15–16.

69 *Twin Peaks*, Creat. David Lynch and Mark Frost (ABC, 1990–91); *Northern Exposure*, Creat. Joshua Brand and John Falsey (CBS, 1990–96).

70 Robert J. Thompson, *Television's Second Golden Age* (New York: Continuum, 1997), 152.

71 Thompson 1997: 155.

72 For discussions of *Twin Peaks* from an aesthetic and/or auteurist perspective, see Brian Jarvis, 'Cherry Pie Heaven: David Lynch', in *Postmodern Cartographies* (London: Pluto Press, 1998); John Alexander, *The Films of David Lynch* (London: Letts, 1993); K. Kaleta, *David Lynch* (New York: Twayne, 1993); David Lavery (ed.), *Full of Secrets: Critical Approaches to Twin Peaks* (Detroit: Wayne State University Press: 1995); Erica Sheen and Annette Davison (eds), *The Cinema of David Lynch: American Dreams, Nightmare Visions* (London: Wallflower Press, 2004).

73 Thompson 1997: 152–3.

74 Key 1980s examples of 'quality television' would be *Hill Street Blues, L.A. Law, Cagney & Lacey, thirtysomething, China Beach, Moonlighting, St Elsewhere, Miami Vice, Shannon's Deal*, and *Cheers*. More recent examples from the 1990s would include the Seattle-set *Frasier* and *Millennium, Picket Fences, NYPD Blue, Homicide: Life on the Street, Ally McBeal, The West Wing, The Sopranos, Six Feet Under* and *The Practice*.

75 Jane Feuer, *MTM: Quality Television* (London: BFI, 1984).

76 Thompson 1997: 14.

77 Smith 1995: 35.

78 Jarvis 1998: 180.

79 Ibid.

80 J. Kingston Pierce, 'Twin Peaks and Other Mysteries', in *Insight Guides: Seattle*, 75–6.

81 Thompson 1997: 164.

82 Smith 1995: 43.

83 Indeed, *Twin Peaks* has been the subject of scholarly essay collections in and of itself. See David Lavery (ed.), *Full of Secrets: Critical Approaches to Twin Peaks* (Detroit: Wayne State University Press: 1995).

84 *Eddie Bauer online*, 'Company background', 5 May 2000, www.eddiebauer.com.

85 Ibid.

86 Ibid.

87 *REI online*, 'Company overview', 12 May 2000, www.rei.com.

88 Ibid.

89 *REI online*, 'About REI'. See also Timothy Egan 1988: 4. There are a number of interesting – and telling – differences between the versions of REI's genesis provided by Egan's article and the REI website. In Egan's interview Lloyd Anderson recalls ordering an Austrian ice-axe, and instead getting a "poor-grade Japanese one". Ten years on, in 1998, the REI website, also containing an article on the new flagship Tokyo REI store, omits the reference to the national origin of the "bad axe", for obvious reasons.

90 *Eddie Bauer online*; *REI online*.

91 Bell and Valentine 1997: 147.

92 Japenga 1992: E8.

93 *Eddie Bauer online*, 'Company History'.

94 'Emerging Markets for Outdoor Recreation in the United States', *ORCA website*, 12 June 2000, www.outdoorlink.com/infosource/nsre/chap2/CH2.htm. The survey was conducted for the Sporting Goods Manufacturers Association (SGMA) and the USDA Forest Service. The report found that since 1982, hiking had an increase in participation of 93.5 per cent; backpacking, 73 per cent; walking for recreation, 42 per cent, from 94 million to 134 million; downhill skiers rose from under 11 million to almost 17 million. The chart excluded new activities, such as snowboarding, that would not have registered on the earlier survey.

95 'Emerging Markets for Outdoor Recreation in the United States'. ORCA also served to corroborate Alexander Wilson's history of recreation nature, stating that "a hundred years ago, Americans realised that their Nation's rapidly expanding industrial systems were not providing satisfactory lives for factory workers." The report concluded that in response to that situation, developments in outdoor recreational opportunities over the last one hundred years have "made impressive improvements in the quality of the lives of American workers".

96 OCRA's report into outdoor recreational activities (produced for industry consumption) identified a market segment that it calls "the enthusiasts [that] comprise the most active one-third of participants in human-powered outdoor recreation." The report stated "Enthusiasts account for a large majority of outdoor recreation participation, even though their numbers are relatively small. Most outdoor recreation enthusiasts, and human-powered enthusiasts in particular, are either under 25 years of age or in their 30s. They are well educated … most of them are men." (ORCA).

97 *Eddie Bauer online*, 'Seattle's Best Coffee and Eddie Bauer Unveil Proprietary Coffee Blend', 27 October 1997, www.eddiebauer.com/about/frame_971027.asp.

98 *REI online*, 'REI Flagship Stores in Tokyo (Japan) and Denver to include Starbucks Coffee Stores', 18 May 2000, www.rei.com/press/starbucks.html.

99 As an aside, it is worth noting that there are a number of examples of this construction of 'Northwest outdoors people' being mobilised within the local Seattle press. For example, in March 1988 Timothy Egan interviewed REI founders Lloyd and Mary Anderson for an article in the *Seattle Times* to coincide with REI's fiftieth anniversary. In the article Egan opined that REI "didn't just grow in Seattle by accident". Adding that "it's hard to imagine Dallas spawning an out-door co-op", Egan linked REI's emergence in Seattle to a populace with a specific constitution, stating that: "In the Northwest, unlike the aggressive

East or the common-sense Midwest or the flakey frontier of Southern California, people are defined more by what they do after work than what they do on the job. There is no 'power city' here – where the Type A can run amok, but there are power winds that whip up the Columbia Gorge." On one level this was fairly innocuous stuff: journalistic hyperbole used to entertaining effect. Yet it did serve to privilege certain constructions of identity. To be a real Seattleite and a real Northwesterner (and not a Type A, running amok) required not only a 'relationship with nature', but also a particularly intimate and demanding connection.

100 Mike Crang, *Cultural Geography* (London: Routledge, 1998), 174–5.
101 Richard Dyer, *White* (London: Routledge, 1997), 21.

CHAPTER 3

1 Actor David Chapelle, playing himself on 'Larry's Sitcom', *The Larry Sanders Show*, HBO, Season Four, Episode 59, BBC 2, 23 February 1999.

2 Much important scholarly work has been undertaken to uncover the very real historical biases and stereotyping at work in the cinematic and televisual (under)representation of African Americans, particularly J. Fred MacDonald, *Blacks and White TV* (Chicago: Nelson-Hall, 1992); Herman Gray, *Watching Race: Television and the Struggle for Blackness*, (Minneapolis: University of Minneapolis Press, 1995); and Ed Guerrero, *Framing Blackness: The African American Image in Film* (Philadelphia: Temple University Press, 1993). In addition, Darrell Hamamoto, *Monitored Peril: Asian Americans and the Politics of TV Representation* (Minneapolis: University of Minnesota Press, 1994), has served to illuminate the comparatively neglected history and politics of the vestiges of Asian American representation in media culture.

3 1990 US Census Data, Database C90STF1A, 18 May 1998, www.venus.census.gov/cdrom/lookup.

4 Stanley Young, 'Beginnings', in John Wilcock (ed.), *Insight Guides: Seattle* (Singapore: APA Publications, 1997), 41.

5 N. W. Barcus, 'Seattle's Jimi', *Stranger*, 12 June 1996. Hendrix in particular is on record as having "little fondness for Seattle where he grew up in poverty". In addition, the mid-1990s witnessed some questioning of the creation of a statue of Hendrix in a predominantly white area of the city, in part due to the alleged lack of support or pride shown by Seattle towards the musician whilst he was alive. Clark Humphrey retells a famous story about Hendrix visiting his alma mater, Garfield High, in 1968. Hendrix was asked when was the last time he had been to Seattle. Humphrey writes "accounts of his response vary, but generally run along the lines of 'About a thousand years ago.'" (1993: 17).

6 Moreover, as Ronald Takaki notes in *Strangers from a Different Shore* (London: Penguin, 1989), given the limited scope in roles granted to Asian Americans within Hollywood casting has tended to oscillate between martial arts movies (*The Karate Kid* (1984); *Rumble in the Bronx* (1995), etc), and signified Chinatown Asian 'mafia' movies (*Year of the Dragon* (1985)), with a few notable exceptions, such as *Living on Tokyo Time* (1987) and Ang Lee's *The Wedding Banquet* (1993). It is worth remembering that such movies invariably gravitate towards New York or Los Angeles, where, as Gina Marchetti suggests in *Romance and the 'Yellow Peril: Race, Sex and Discursive Strategies in Hollywood Fiction* (Berkeley: University of California Press, 1993), well-known, tourist-enticing, densely-populated Chinatowns allow

"pure style with neon dragons, pop songs, lion dances, and displays of martial artistry" to "play with the dangers of the exotic" (34–8). Seattle, although home to two of the most significant Asian American performers does not possess the required urban aesthetic to fit this remit, failing to possess the 'colourful' Chinatown (instead, a low-rise, functional 'International District') to attract such attention. Although born in San Francisco, Bruce Lee grew up in Seattle and attended the University of Washington. He is buried in the city's Lake View cemetery, alongside Brandon Lee, his son. Keye Luke (1904–91) was born in Canton, China, but raised in Seattle. His film debut was in *The Painted Veil* (1934) with Greta Garbo (see Hamamoto 1994: 238).

7   David A. Takami, *Divided Destiny: A History of Japanese Americans in Seattle* (Seattle: University of Washington Press, 1998), 18; Roger Sale, *Seattle, Past to Present* (Seattle: University of Washington Press, 1976), 37–49.

8   Quintard Taylor points out that in the nineteenth century the Pacific Northwest was "often described by land promoters and politicians as 'the white man's country'" (1994: 23).

9   John M. Findlay, 'Regional Identity in the Pacific Northwest', in Wrobel & Steiner (eds) 1997: 46. Of particular note are the anti-Chinese riots in 1886, when Chinese homes and stores were demolished, and 200 Seattle Chinese were forced to board a steamship bound for San Francisco. See Roger Sale, *Seattle, Past to Present* (Seattle: University of Washington Press, 1976), 37–49. In 1942, 6,000 members of the Seattle Japanese community were rounded up and shipped to an internment camp in Idaho. David Guterson's novel *Snow Falling on Cedars*, (London: Bloomsbury, 1995) and Monica Sone's *Nisei Daughter* (Washington: University of Washington Press, 2003) both provide interesting accounts of the Japanese population during this period.

10 Findlay 1997. Incidentally, the *Seattle Times*, 19 March 1999, carried a piece by Janet Burkitt entitled 'Hate mail targets interracial couples', cataloguing an increased spate of such postings in the Seattle area.

11 Mary Bruno, 'Seattle Under Seige', *Lear's*, July 1991: 53; Jerry Adler, 'Seattle Reigns', *Newsweek*, 26 May 1996: 48; Dean Paton, 'Seattle's rise as the capital of the New Economy', *Christian Science Monitor*, 30 November 1999: 1–4.

12 It is worth noting how rapidly this process seems to have occurred. For example, Jonathan Raban, in the chapter 'Gold Mountain' from his collection of travel writings entitled *Hunting Mr Heartbreak* (London: Collins Harvill, 1990) claimed that "you could find Seattle in novels, but there wasn't a Seattle Novel" (357). He stated that "a truthful Seattle novel would have to be about the newcomers", and concluded the chapter with his own tentative synopsis for a Seattle Novel, which cast a young white teacher in a romance with his divorced Korean student, both new arrivals to the city. Perhaps Raban would have had less hesitancy in proclaiming the existence of a Seattle Novel at the end of the 1990s, given the growth of writers who have sought to depict 'the newcomers'.

13 Stuart Hall, 'The Question of Cultural Identity', in Stuart Hall, David Held and Tony McGrew (eds) *Modernity and Its Futures* (Cambridge: Polity Press, 1992), 309. It is also worth mentioning that the city's detective fiction writers have also drawn upon Seattle's multiculturalism in order to add local nuance to their take on the genre. In J. A. Jance's *Dismissed With Prejudice* (New York: Avon Mystery, 1989) Detective J. P. Beaumont investigates the 'ritual' murder of a Japanese American businessman, and includes a brief commentary on the World War Two Japanese relocation camps in the Pacific Northwest.

Earl Emerson's *Catfish Café* (New York: Ballantine Books, 1998) weaves (albeit a rather mawkish) social commentary on the plight of impoverished African Americans in Seattle's Central District into private eye Thomas Black's search for a missing woman, whilst in Louise Hendricksen's *Lethal Legacy* (New York: Zebra Books, 1995) Dr Amy Prescott searches for a murderer at large in the area's Cambodian community. These authors, invariably white, could be criticised for turning the region's non-white inhabitants into a roster of corpses, murderers, alibis and minor players – it is worth noting that none of Seattle's fictional detectives are people of colour. However, in at least acknowledging the cultural diversity of the Puget Sound area, detective fiction seeks to keep apace with demographic transformations in the real city.

14 Homi Bhabha, 'Novel Metropolis', *New Statesman and Society*, 16 February 1990: 16.

15 Details from Alex Tizon, 'Peter Bacho's world', *Seattle Times*, 1 March 1998.

16 Hall, 306.

17 James Donald, 'Metropolis: The City as Text', in Robert Bocock, Robert Thompson and Kenneth Thompson (eds), *Social and Cultural Forms of Modernity* (Cambridge: Polity Press, 1992), 456.

18 Doreen Massey, 'Power-geometry and a progressive sense of place', in Jon Bird, Barry Curtis, Tim Putnam, George Robertson and Lisa Tickner (eds), *Mapping the Futures* (London: Routledge, 1993), 60.

19 Davis 1990: 224–8.

20 Sharon Zukin, *The Culture of Cities* (Oxford: Blackwell, 1995), 26.

21 Zukin 1995: 43.

22 Donald 1992: 456. Indeed, a not dissimilar conclusion was reached by Paul Gilroy in *There Ain't No Black in the Union Jack* (London: Hutchinson, 1987). Gilroy noted that the British media tended to mobilise race as a synonym for urban crisis (228–9).

23 Beauregard 1993: 173; 290.

24 As Beauregard states, "racism is profoundly embedded in white attitudes and behaviour thereby making racial identity a symbolic magnet that attracts the myriad problems associated with the cities. The disentangling of the contributions of racial discrimination and minority behaviour from the many forces shaping urban decline in objective, much less discursive, terms thus becomes an interpretative quagmire. Within the discourse, urban decline is a virtual stand-in for race and the two are frequently indistinguishable" (1993: 291).

25 Beauregard 1993. He writes that "The Negro was situated at the core of physical deterioration, white flight, anaemic capital investment, crime, poverty, poor schools and unemployment. Binding that core together was fear, fear that centuries of racism and inequality would finally accumulate in insurrection. With race as the dominant discursive thread, the various fragments of urban decline became whole cloth" (164).

26 Beauregard 1993: 219.

27 Beauregard 1993: 266; 267.

28 Hamamoto 1994: 170; 236. Although, as Hamamoto recounts, one famous image showed "Korean-American merchants – one wearing a bullet-proof vest – firing their weapons at attackers … [and] will leave a lasting impression in the mind of the public and might even begin to alter the manner in which Asian Americans are portrayed on Network television."

29 Michael Medved, 'Paradise Lost – Riots, Fire, Slump and now Earthquake – for many Californians the dream is over', *Los Angeles Times* (reprinted in *The Sunday Times*, 23 January 1994).

30 Martin Walker, 'Washington State replaces California as haven for infrastructure investment', *Guardian*, 14 September 1992. Reuters News Agency.

31 Joel Kotkin, 'The Danger of the California Exodus Myth', *Seattle Times*, 10 March 1993: A7.

32 Ibid.

33 Davis 1999: 371.

34 Kennedy 2000: 7. Interestingly, the Mayor overseeing Seattle from the late 1980s onwards was Norm Rice, the first African-American Mayor to be elected in the city. Whilst the local press in Seattle attributed the city's relative calm during the L.A. riots to the fact that "[Rice] consulted black community leaders, took their advice and stayed in control", representations of Rice on a national and international scale paid much more attention to the fact that he was the black mayor of a 'white city'. ('Rice's legacy: downtown and caring stewardship', *Seattle Times*, 28 December 1997). Rice became mayor of Seattle in 1989, a year that also saw David Dinkins become the first black mayor elected in New York. Whilst Dinkins became the head of a city widely understood as the world's most ethnically diverse metropolis, Norm Rice was described internationally as "clean[ing] up a white city" (John Carlin, 'Seattle's first black mayor cleans up a white city', *Independent*, 15 July 1996).

35 Kennedy 2000: 17–18.

36 Richard Dyer, 'White', *Screen* 28, 4, 1988: 44–4.

37 Richard Dyer, *White* (London: Routledge, 1997), 1.

38 Dyer 1997: 3.

39 Ibid.

40 John Gabriel, *Whitewash: Racialized Politics and the Media* (London: Routledge, 1998), 14.

41 Dyer 1997: 4.

42 bell hooks, *Black Looks* (Boston MA: South End Press, 1992); Fred Pfeil, *White Guys* (London: Verso, 1995); Dyer 1997: 4.

43 Gabriel 1998: 2.

44 Dyer 1997: 16.

45 Gabriel 1998: 19; 17; 2.

46 Dyer 1997: 217–22. See also Jude Davies, '"I'm the Bad Guy?": *Falling Down* and White Masculinity in 1990s Hollywood', *Journal of Gender Studies* 4, 2: 145–52; Liam Kennedy, 'Alien Nation: White Male Paranoia and Imperial Culture in the United States', *Journal of American Studies*, 1996, 30, 1: 87–100; Pfeil 1995: 238–43; John Gabriel, *Whitewash* (London: Routledge, 1998), 134.

47 Rita Kempley, 'The Hand That Rocked The Cradle', *Washington Post*, 10 January 1992; Review of *The Hand That Rocked The Cradle, Rolling Stone*, February 1992.

48 Dyer 1997 19.

49 Kotkin 1993.

50 Details from Chuck Taylor, 'Seattle's a bit player', *Seattle Times*, 27 April 1997: E12. Entering into its ninth season in 2001, *Frasier* had earned a total of 21 Emmy awards, and has also made history by becoming the first show to win five consecutive Emmy awards for Outstanding Comedy Series.

51 Robert J. Thompson, *Television's Second Golden Age* (New York: Continuum, 1997), 14.

52 See, for example, Jane Feuer, Paul Kerr and Tise Vahimagi (eds), *MTM 'Quality Television'* (London: BFI, 1984); Mark Jancovich and James Lyons (eds) *Quality Popular Television*

(London: BFI, 2003).

53 Peter Casey, quoted in Chuck Taylor, 'Seattle's a bit player'. This notion is discussed in more detail in chapter six.

54 My thanks to Jeffrey Miller for his comments and suggestions in relation to this notion.

55 'Merry Christmas, Mrs Moskowitz', *Frasier*, NBC, Series 6, Channel 4.

56 Many of the creative team behind *Frasier* were also responsible for *The Mary Tyler Moore Show* which ran on CBS between 1970 and 1977, and set itself in another visually 'white' city, namely Minneapolis. As such, it could be possible to suggest, very tentatively, that different cities may function in the media as fantasy liberal white metropoles in different periods, although evidence for such an assertion would require research broader than this chapter affords.

57 Susan Gubar, *Racechanges: White Skin, Black Face in American Culture* (Oxford: Oxford University Press, 1997), xiii.

58 Gubar 1997: xiv.

59 Gubar 1997: xv.

60 'Something about Dr. Mary', *Frasier*, NBC, Series 7, Channel 4.

61 Gubar 1997: xix.

62 Gubar 1997: 246.

63 Gubar 1997: 241.

64 Stuart Hall, 'What is this "Black" in Black Popular Culture?', in Gina Dent (ed.), *Black Popular Culture, a Project by Michelle Wallace* (Seattle: Bay Press, 1992), 21–33.

65 Gubar 1997: 241.

66 Gubar 1997: xviii.

67 Kennedy 2000: 7.

CHAPTER 4

1 Frederick Jackson Turner, from 'The West and American Ideals', in Carl Abbott, *The Metropolitan Frontier* (Tuscon: University of Arizona Press, 1993), 179. It is interesting to note that from early in its history, Seattle has been depicted as a place that could provide a better environment in which to raise a family. Roger Sale, in his *Seattle, Past to Present* (Seattle University of Washington Press, 1976) repeats the traditional account of the founding of Seattle. It was a markedly family affair; Sale writes that the schooner *Exact*, having travelled up from Portland, "stopped at what is now called Alki Point in West Seattle. Twenty-two people, ten adults and twelve children disembarked" (7).

2 J. A. Jance, *Without Due Process* (New York: Avon Books, 1992), 68.

3 Neil Smith, *The New Urban Frontier* (London: Routledge, 1996), xiv.

4 Beauregard 1993: 257.

5 Neil Smith and Peter Williams, 'Alternatives to orthodoxy: invitation to a debate', in Neil Smith and Peter Williams (eds), *Gentrification of the City* (London: Unwin Hyman, 1988), 1–3.

6 For a detailed discussion of this, see Robert Beauregard, "Chaos and Complexity in Gentrification," in *Gentrification of the City*.

7 Beauregard 1993: 119–121.

8 See Joel Garreau, *Edge City* (New York: Doubleday, 1991).

9 Beauregard 1993: 307.

10 Details from Arthur M. Louis, 'The Worst American City' in *Harper's*, January 1975: 67–104. Mencken came to the conclusion that the worst state was Mississippi, the best Massachusetts.

11 Louis 1975: 67.

12 Ibid.

13 Beauregard 1993: 225.

14 Beauregard 1993: 198. Quote from Commission on the Cities, *The State of the Cities* (New York: Praeger, 1972), 5.

15 Beauregard 1993: 71.

16 Sale notes that between 1960 and 1970 Seattle "went from last to first … among west coast ports in shipping to the Overland Common Points farther east" (1994: 234).

17 Norbert MacDonald, *Distant Neighbors* (Lincoln: University of Nebraska Press, 1987), 182.

18 Sale 1994: 239.

19 Ibid.

20 Beauregard 1993: 247.

21 David Harvey, *The Condition of Postmodernity* (Oxford: Blackwell, 1989), 147.

22 Beauregard 1993: 246. As long as the rich got richer quicker than the poor got poorer, the average wealth rating of a city's populace would rise.

23 'Ranking the Cities', *Time*, 29 September 1975: 39–40. See also, William Marlin and Roland Gelatt, 'America's most livable cities', *Saturday Review* 3, 21 August 1976.

24 Judith Frutig, 'America's Most Livable Cities', *Christian Science Monitor*, 21 May 1975: 27–32.

25 Frutig 1975: 27.

26 Ibid.

27 Ibid.

28 Whilst this may reflect a notion of a 'new level' of consumer culture, its seductiveness could also be said to be to lie in the fact that it is in keeping with an urge that has characterised the New World since inception – namely for setting off for new territory, and heading west.

29 In 1977 *Town and Country*, a magazine focusing upon the lifestyles of America's upper classes, declared Seattle "a city with big-time sophistication and small-town accessibility". Reprinted in Robert T. Nelson, 'One more organisation discovers Seattle can be a great place to live', *Seattle Times*, 13 November 1986, Northwest Section: F4.

30 'Dixy Rocks the Northwest', *Time*, 12 December 1977: 26–36.

31 Beauregard 1993: 247.

32 See, for example, Blake Fleetwood, 'The new elite and an urban renaissance', *New York Times Magazine*, 14 January 1979, or Penelope Lemov, 'Celebrating the city', in *Builder*, 7, February 1984: 90–7.

33 Harvey 1989: 92.

34 Beauregard 1993: 256.

35 Mary Ellen Mark and Cheryl McCall, 'Streets of the Lost', in *Life*, July 1983: 34–42.

36 Mary Ellen Mark achieved worldwide visibility through numerous books (for example, *Passport* (1974); *Ward 81* (1979); *Falkland Road* (1981); *Mother Teresa's Mission of Charity in Calcutta* (1985); *The Photo Essay: Photographers at Work* (1990); *Streetwise* (1992); *Mary Ellen Mark: 25 Years* (1991); *Indian Circus*, (1993) *Portraits* (1995); and *a Cry for Help* (1996)). She published photo-essays for *Life*, *New York Times Magazine*, *New Yorker*, *Rolling Stone* and *Vogue*. Her

professional recognition includes the John Simon Guggenheim Fellowship, the Matrix Award for outstanding woman in the field of film/photography, and the Dr Erich Salomon Award for outstanding merits in the field of journalistic photography, Photographer of the Year Award from the Friends of Photography; the World Press Award for Outstanding Body of Work Throughout the Years; the Victor Hasselblad Cover Award; two Robert F. Kennedy Awards; and the Creative Arts Award Citation for Photography at Brandeis University.

37   Mark and McCall 1983: 34.
38   Mark and McCall 1983: 35.
39   Pileggi 2000: 35.
40   Pileggi 2000: 35, 36, 42.
41   Indeed, from as early as the 1950s onwards, commentators noted that the economic fortunes of cities were being effected seriously by large numbers of the middle class – "particularly the families with small children" – finding its way to suburbia, and in the process, being seen to deprive the cities of "taxpayers, workers and consumers" (Beauregard 1993: 124).
42   Reprinted in Stephanie Coontz, *The Way We Never Were* (New York: Basic Books, 1992), 256.
43   Arlene Skolnick, *Embattled Paradise: The American Family in an Age of Uncertainty* (New York: Basic Books, 1991), 6.
44   Ibid.
45   Ibid.
46   Kennedy 2000: 93–4.
47   *Streetwise*, dir. Martin Bell, Angelica Films/Bear Creek, 1984.
48   Kennedy 2000: 99.
49   *American Heart*, dir. Martin Bell, perf. Jeff Bridges and Edward Furlong, Triton Pictures, 1992.
50   Box-office details from the *Internet Movie Database*, www.imdb.com. The British film magazine *Sight and Sound* praised the performances of Bridges and Furlong, and applauded what it saw as the movie's unwillingness to imbue the characters' "low-rent life-style" with either sentimentality or "sleazy kitsch glamour" (Lizzie Francke, review of *American Heart*, *Sight and Sound*, May 1992: 40–1). Similarly, *Empire* magazine made positive remarks about Bridges' performance, whilst *Rolling Stone* and *The New York Daily News* both gave the movie positive reviews.
51   *Sleepless in Seattle*, dir. Nora Ephron, perf. Tom Hanks and Meg Ryan, Tri-Star/Sony,1993.
52   Something also underlined by Maggie's ghostly apparition in the movie – depicted in an angelic white light.
53   Thomas Schatz, 'The New Hollywood', in Jim Collins. Hilary Radner and Ava Preacher Collins (eds), *Film Theory Goes to the Movies: Cultural Analysis of Contemporary Film* (New York: Routledge, 1993), 19.
54   Emmett Watson, 'Most Livable City? It's Time to Berate the Ratings', *Seattle Times*, 26 October 1989, Northwest section: C1. Watson wrote complaining about being contacted by journalists from CBS and *USA Today* for his response to Seattle's number-one ranking.
55   *Savvy* Magazine, quoted in Richard Seven, '"Most Livable City" – Seattle hits top of the metropolitan ratings', *Seattle Times*, 25 October 1989, Northwest section: D1; *Money* August 1989. As an example of just how seriously the ratings charts were taken, in October 1989 Seattle mayor Charles Royer attended a ceremony at the offices of *Places Rated Almanac*

publisher Prentice Hall in New York. The mayor of Pittsburgh, the previous 'most liveable city', handed Seattle mayor Charles Royer a commemorative silver platter. Details from Frederick Case, as above.

56 Nora Ephron, in interview with Lawrence Frascella, *Rolling Stone*, 8 July 1993, 73–5.

57 Robins 1993: 306.

58 *The Stepfather*, dir. Joseph Ruben, perf. Terry O'Quinn and Shelly Hack, New Century/ Vista, 1987. Indeed, *The Hand That Rocks The Cradle* could be argued to display evidence of the *The Stepfather*'s influence.

59 It should be noted that the film *Copland* and the HBO television series *The Sopranos* may have done much to reinforce this sentiment.

60 'Meet Frank Black, Everypatriarch, on a mission to keep ugliness from tainting his family', *Time*, 28 October 1996: 100.

61 Episode Guide, "Pilot", *Millennium* official web site, 7 September 1999, www.foxnetwork.com/ millnium/epi100.htm.

CHAPTER 5

1 David Morley and Kevin Robins, 'Reimagined Communities?', *Spaces of Identity* (London: Routledge, 1995), 37.

2 Caption from inside cover of *Hype!* The Motion Picture Soundtrack, Sub Pop Records, 1996.

3 Whilst it is true that radio and MTV are also important disseminators of information concerning popular music, they are not included here mainly because of the impossibility of getting access to archived material. Nevertheless, an examination of the print media does provide an essential, and consistent, source of information on popular music, and remains crucial to the generation of popular music discourses.

4 Hall 1992: 304.

5 Zukin 1995 20–3.

6 See Ronan Paddison, 'City Marketing, Image Reconstruction and Urban Regeneration', *Urban Studies*, 30, 2, 1993: 339–50.

7 Details from Christopher Sanford, *Kurt Cobain* (London: Vista, 1996), 372.

8 Clark Humphrey, *Loser: The Real Seattle Music Story* (Portland: Feral House, 1995), 171.

9 Dave DiMartino, 'A Seattle Slew', *Rolling Stone*, 20 September 1990: 23. Minneapolis had produced The Replacements and Husker Dü, whilst Athens, Georgia was the birthplace of REM and the B52s, amongst others.

10 See for example, Greil Marcus, *Mystery Train: Images of America in Rock 'n' Roll Music* (New York: Plume Books, 1997).

11 Will Straw, 'Systems of Articulation, Logics of Change: Communities and Scenes in Popular Music', *Cultural Studies*, 5, 3, October 1991: 373. Straw is drawing upon 'Transgressing the boundaries of a rock 'n' roll community', a paper Barry Shank delivered at the *First Joint Conference of IASPM-Canada and IASPM-USA*, Yale University, 1 October 1988.

12 Straw 1988: 373.

13 Straw 1988: 389.

14 Straw 1988.

15 Barry Shank, *Dissonant Identities: The Rock 'n' Roll Scene in Austin, Texas* (Connecticut:

Wesleyan University Press, 1994), 122.

16 Sanford 1996: 81–3.

17 DiMartino 1990: 23.

18 Brad Morrell, *Nirvana and The Sound of Seattle* (London: Omnibus Press, 1996), 14. Such claims for isolation, it must be stated, did possess some empirical validity, at least in terms of Seattle's position within the national and international networking of the music industry. As Clark Humphrey points out, coming into the 1980s, Seattle was isolated "from the music industry. The Big Six labels and their subsidiaries had no A&R people in the North-west states." This is a view reiterated by local record producer and former band member Jack Endino, who stated that "when you were living in Seattle – you know, it wasn't LA – nobody was going to come and sign us" (Humphrey 1995: viii.)

19 Kevin Robins, 'Tradition and translation: national culture in its global context', in John Corner and Sylvia Harvey (eds), *Enterprise and Heritage: Crosscurrents of National Culture* (London: Routledge, 1991), 29; Hall 1992: 304.

20 Inga Moscio, 'Mister Sub Pop: Jonathan Poneman', *Stranger*, 27 September 1993: 3.

21 Humphrey 1995: 49.

22 Humphrey 1995.

23 Humphrey 1995: 35, 125. Poneman had himself been a guitarist with a number of local bands from the early 1980s onwards.

24 Humphrey 1995: 138.

25 Ibid.

26 Interviewed for *Hype!*

27 Humphrey 1995: 133.

28 Interviewed for *Hype!*

29 Humphrey 1995: 140. The suggestion of "no women in sight" was particularly interesting, given that Pavitt had relocated from Olympia, a city that would subsequently become synonymous with the 'Riot Grrrls' movement of "young feminist women in underground rock". Indeed, as Joanne Gottlieb and Gayle Wald point out (in 'Smells Like Teen Spirit: Riot Grrrls, Revolution and Women in Independent Rock', in Andrew Ross and Tricia Rose (eds), *Microphone Fiends: Youth Music and Youth Culture* (London: Routledge, 1994), 250) Nirvana's ground-breaking *Smells Like Teen Spirit* was an appropriation by Nirvana's Kurt Cobain of graffiti scrawled on the wall of his Olympia House by Kathleen Hanna, a member of the Riot Grrrl band Bikini Kill. As Gottlieb and Wald note wryly, the incident "hint[ed] at the creative invisibility of a woman behind what was to become a ubiquitous, industry-changing, Top 10 hit for a male rock group". Although Riot Grrrl bands would see their own move into the limelight and "'mainstream' mass market venues" in 1992, as major label deals for the bands Babes in Toyland, L7 and Hole (fronted by Cobain's wife Courtney Love) took effect, Sub Pop's marketing strategy served to give the 'Seattle sound' a male vocal accompaniment (Ibid.).

30 Humphrey 1995: 135.

31 Humphrey 1995: 136.

32 Everett True, 'Sub Pop, Sub Normal, Subversion: Mudhoney', *Melody Maker*, 11 March 1989: 28–9. As an interesting aside, True would eventually move to Seattle in 1998, to become a music critic for the *Stranger*.

33 Concerns with notions of 'unique', unformulaic, and thus 'authentic' music are, on one

level, merely the reiteration of a well-worn theme within rock music's own paradigmatic narrative framework, namely the signified 'struggle' against the co-option of 'the real' and 'the authentic' music by what is conceived to be the commercial mainstream. The character of this narrative has been explored usefully by Lawrence Grossberg, who argues that what he terms "the operational logic [of] the rock formation" supports, and has supported historically, a rhetoric of its own death. As Grossberg states in 'Is Anybody Listening? Does Anybody Care? On "The State of Rock"', in *Dancing in Spite of Myself* (North Carolina: Duke University Press: 1997, 102–21), "this rhetoric assumes a binary, hierarchical and cyclical map of the musical terrain. It is built upon a strategy of differentiation: always distinguishing between the authentic and the co-opted. This distinction, then, easily and often slides into a narrative war between authentic youth cultures and corrupting commercial interests. Rock is judged dead to the extent that the commercial interests, the co-opted music, seem to be in control, not only of the market, but of the music and the fans as well" (103). Grossberg concludes that within the logic of this rhetoric, "rock is never dead, but it is constantly in the process of dying or of being killed". It is not difficult to find popular accounts of grunge music's commercial ascendancy that supported this rhetoric. For example, in *Nirvana and the Grunge Revolution* (Millwaukee: Hal Leonard, 1998), Jeff Kitts, Brad Tolinski and Harold Steinblatt provide a not untypical example. The authors state that one subgenre of rock, namely pop-metal, had dominated the music charts and MTV in the mid- to late 1980s. Represented by bands such as Bon Jovi, Warrant, Poison and Motley Crue, pop-metal was being castigated in the rock press, with the support of the authors, for being "pouffed hair, simpering ballads … every song pulsated with explosive but empty guitar solos … [it was] a neon carnival filled with pomp". As Sub Pop's Jonathan Poneman asserted, the "world of rock [had become] increasingly safe and bloodless" ('Grunge and Glory', *Vogue*, December 1992: 260). Grunge, as Javier Santiago-Lucerna notes in his essay '"Frances Farmer will have her revenge on Seattle": Pan-capitalism and Alternative Rock', in Jonathan S. Epstein (ed.), *Youth Culture: Identity in a Postmodern World* (Oxford: Blackwell, 1999), 189–90, was by contrast identified as "a strong statement against the glam aesthetic of eighties heavy metal … a return to the emotionally charged performance, the display of raw power as an assurance of the truthfulness and sincerity of the performer[s]". Thus the narrative of grunge music's ascendancy is characterised by a struggle over authenticity: pop-metal had 'sold out' to corporate interests, and no longer connected on an affective level with American youth. In this way grunge's supplanting of pop-metal is predicated precisely upon a binary opposition, and operates within a strategy of differentiation. In positioning itself in relation to struggles of authenticity, narratives of grunge music locate it quite comfortably within rock's lineage. As Grossberg states, "the history of rock is marked by a continuous struggle over what is really authentic rock and which groups are really invested in it" (113).

34 True 1989.

35 Straw 1988: 371.

36 Straw 1988.

37 Ibid.

38 Humphrey 1995: 140.

39 Straw 1988: 375.

40 Straw 1988.

41 Straw 1988: 376.

42 Straw 1988: 378.

43 This line of thinking finds support by Lawrence Grossberg in his writing on the changing status of rock music in the 1990s. In the aforementioned *Dancing in Spite of Myself*, Grossberg discusses the emergence of rock within the particular social and economic context of the post-war years, and the ways in which rock emerged from, and responded to, the emergence of a large youth population. Grossberg asserts that the 1990s witnessed a distinct change in "the nature and effects of popular music". On one level, this is an argument about the transformation of the media economy, and the influence of new technology. Grossberg states that for the first time "the visual (whether MTV or youth films or even network television) … is increasingly displacing the aural as the locus of generational identification, differentiation, investment, and occasionally even authenticity". Grossberg claims that the 1990s evidenced the role of popular music being transformed, and serving increasingly as merely the soundtrack for other affective media. It is a provocative and contentious claim, but one that did capture something of the changing media landscape, particularly the growing influence of multi-media technology and the ascendancy of computer game culture (which possesses its own charts, heroes and anthems). In addition, it also included a crucial recognition of the different affective relationship to music engendered by the burgeoning area of dance music (house, garage, drum and bass, and so forth). This was captured neatly in Grossberg's notion that in the 1990s what mattered less and less was dancing to the music you liked, than "liking the music you can dance to" (115–20).

44 Dick Hebdige, *Subculture: The Meaning of Style* (London: Routledge, 1989), 92–3.

45 Hebdige 1989: 95.

46 Clip from *Hype!*

47 For many observers in Seattle this was an example of the previously cited fears of grunge's co-option into the mainstream taken to its absurdist extremes, and into realms of misappropriation previously unimaginable. In the *Hype!* documentary, Susan Silver, manager of the band Soundgarden stated that the East Coast fashion spreads by *Vogue* and *Elle* represented for her the only moment of the grunge phenomenon that became "unbearable". So synonymous had the city become with grunge and all things "cool" that, as Chris Eckman of the Seattle band The Walkabouts stated, perhaps somewhat apocryphally, "In Europe, they just started to put stickers on things that just said Seattle, that's all the sticker said … it's like the USDA: stamp it, 'Seattle'." The difference was that this was not governmental branding, but commercial branding, and Seattle had attained the status to which its civic boosters would no doubt have wished it to aspire – a trademark icon. The label that was being hyped was not Sub Pop, but Seattle itself, and in short, Seattle *sold*.

48 Grace Coddington, 'Grunge and Glory', *Vogue*, December 1992: 254–63.

49 Courtney Love was the wife of the lead singer of Nirvana, the late Kurt Cobain, but also the lead singer of the successful grunge rock band Hole.

50 Hebdige 1989: 93.

51 Crowe was far from unfamiliar with Seattle; he had lived in the city since marrying Nancy Wilson, Seattle-born member of the rock band Heart in 1986. Moreover, he had used Seattle as the location for his first feature film *Say Anything*, in 1989, a humorous examination of American teenage life, featuring songs by local bands Mother Love Bone and Mudhoney on the soundtrack.

52 Dario Scardapane, 'rock auteur', *Vogue*, September 1992: 330–2.

53 Scardapane 1992: 330. As Crowe put it, reflecting upon the "ethical local music" community, "to capture the mentality of people that aren't on the next bus out as soon as they make their first dollar".

54 Details from Morrell 1996: 113.

55 Scardapane 1992: 330.

56 Scardapane 1992: 331.

57 Ibid.

58 As Morrell notes, for those in Seattle who considered themselves close to the actual 'scene', the film was "greeted by purists as a sickening exploitation of the city and its music" (Morrell 1996: 113). The reviewer for the aptly-titled local Seattle weekly *Hype* wrote, "I'm embarrassed by the fact that these guys, along with Matt Dillon, will be representing the Seattle scene to the rest of the nation on the big screen" (Humphrey 1995: 165). In short, the director stood accused of jumping on a mainstream media bandwagon that was nowhere in sight at the time of the film's inception. It could be argued that as a signified 'Hollywood' production, *Singles* would inevitably have incurred the wrath of those who would castigate what they saw as any harnessing of 'the scene' to the furthering of mainstream commercial interests.

59 David M. Gross and Sophfronia Scott, 'Proceeding With Caution', *Time*, 16 July 1990: 57.

60 Ibid.

61 Laura Zinn, 'Move Over Boomers', *Business Week*, 14 December 1992: 74–82; Joseph P. Shapiro, 'Just Fix It!', *US News and World Report*, 22 February 1993: 50–6; Neil Howe and William Strauss, 'The New Generation Gap', *Atlantic Monthly*, December 1992: 67–89. Howe and Strauss also produced a book on the subject, entitled *13th Gen: Abort, Retry, Ignore, Fail?* (New York: Vintage Books, 1993).

62 Howe and Strauss 1992: 69.

63 Gross and Scott 1990: 58.

64 Ibid.

65 Douglas Coupland, *Generation X* (New York: St Martin's Press, 1991). Coupland's novel prompted much discussion in the American newspapers at its time of publication. Its depiction of post-college Americans 'downsizing' and rejecting so-called yuppie values made it easily and eagerly assimilable to the wider discourse on the twentysomethings. See, for example, Robin Abcarian, 'Boomer Backlash', *Los Angeles Times*, 12 July 1991, E: 1; Clarence Page, 'Lament of the twentysomethings', *Chicago Tribune*, 11 August 1991, Section 4: 3.

66 'Marketing Tools', *American Demographics*, Fall 1993: 1; Scott Donaton, 'The Media Wakes Up To Generation X', *Advertising Age*, 64, 5, 1 February 1993: 16–17. See also 'Forget the "X", Please', *Mediaweek*, 2 August 1993: 10.

67 Donaton 1993: 16.

68 It should be stated that Nirvana were not actually from Seattle. Kurt Cobain and guitarist Krist Novoselic were from Aberdeen WA, whilst eventual permanent drummer Dave Grohl was from Virginia. Moreover, the band's original identity was as an Olympia band. This fact says much about the process by which the 'Seattle Scene's' representation in magazines tended to omit such troublesome particulars.

69 Details from Humphrey 1995: 172.

70 This period arguably preceding the instigation of the dance superclubs, and star DJs, as well as DJ albums, that have served to move dance music much closer to the traditional accoutrements of rock.

71 Christopher John Farley, 'Rock's Anxious Rebels', *Time*, 25 October 1993: 60.

72 *Time* was not alone in drawing such epochal conclusions about the band. In a lengthy article for *Rolling Stone* in the same month, movie director and former *Rolling Stone* journalist Cameron Crowe recounted an early *Los Angeles Times* review comparing Pearl Jam's song 'Alive' to The Who's 'My Generation' (Cameron Crowe, 'Five against the world', *Rolling Stone*, 28 October 1993: 56). In addition, Crowe himself commented that 'Today, "Alive" is a Gen X rallying cry.'

73 Farley 1993: 62.

74 It should be noted that Eddie Vedder and Pearl Jam had declined to participate in the *Time* cover story. Vedder later commented that 'This is my parent's magazine, if I had parents', a succinct remark that managed to include the signified rejection of the 'mainstream', to valorise a generation-oriented mode of analysis, and to authenticate Vedder's own identity as a 'damaged' member of 'Generation X' (Humphrey 1995: 194). It is possible to argue that *Time* probably realised from the outset that the band would refuse to assist with the article; this was all part of the same package of constructing the legitimacy of their portrait of 'rock 'n' roll' rebel status. One could have some sympathy with bands caught in a peculiarly postmodern conundrum, whereby any rejection of the mainstream had such a routinised, symbolic quality – and was quite neatly incorporated back into the narrative as the signifier of authenticity, that there seemed to be no space available 'outside'.

75 Farley 1993: 66.

76 For example, Hi-NRG music from 1970s gay disco was being reworked as high-house (Straw 1991: 383).

77 Tricia Rose, 'A Style Nobody Can Deal With: Politics, Style and the Postindustrial City in Hip Hop', in Ross and Rose 1994: 83.

78 Ibid.

79 Rose 1994: 71.

80 Rose 1994: 84

81 Murray Forman, '"Represent": race, space and place in rap music', *Popular Music* (2000), 19, 1: 68.

82 Forman 2000: 71–3.

83 As Foreman states, "since roughly 1987 hip hop culture has also been influenced by alliances associated with West Coast gang systems. Numerous rap album covers and videos feature artists and their posses representing their gang, their regional affiliations or their local 'hood with elaborate hand gestures. The practice escalated to such an extent that, in an effort to dilute the surging territorial aggression, Black Entertainment Television (BET) passed a rule forbidding explicitly gang-related hand signs on its popular video programmes" (2000: 72).

84 As Rose states, "[Hip hop] attempts to negotiate new economic and technological conditions, as well as new patterns of race, class and gender oppression in urban America, by appropriating subway facades, public streets, language, style and sampling technology" (1994: 72.)

85 Farley 1993: 66. It is worth remarking upon a range of post-grunge Hollywood films set in Seattle which drew upon this notion of the city, although mainly to no great success or critical acclaim. The first was *Mad Love* (1995), a romantic drama charting the tempestuous love affair between two high-school students, Matt Leland and Casey Roberts. Casey, newly arrived in Seattle with her parents, was a bleached-blonde 'riot grrrl' who drove a noisy lemon-yellow VW Bug, and was accompanied everywhere by a blasting soundtrack of alternative music,

(in particular by local Seattle band 'Seven Year Bitch'). The story plotted Casey's mental breakdown, evidenced in riotous, iconoclastic actions, her rebelliousness read on the surface as the rejection of a 'boring' city that failed to live up to the promise of a vibrant grunge venue. *Georgia* (1996) was more directly concerned with the local Seattle music scene, focusing on the relationship between two sisters, both singers with strikingly different careers. One sister, Sadie with her thrift-store wardrobe, black eye make-up, drink and drug problems and limited vocal range, was a bar-band singer, playing grunge-tinged neo-punk refrains to sparsely populated venues on the fringe of the local live-music circuit. In contrast, the other sister, Georgia was a popular and very successful country-folk singer, complete with supportive husband, attractive child and beautiful home. The two musical genres functioned as exemplars for the personality types of each sister. Georgia was the epitome of crafted musicianship, with her professionally trained voice, carefully controlled stage performance, and businesslike approach. Sadie, by contrast, was as chaotic and anarchic onstage as off. *Georgia* was ripe for interpreting along iconographic lines – punk versus folk equals 'damaged' Generation X against "the baby boom middle class". The film also garnered notable attention in the local Seattle press, as reviewers expressed interest in a film which purported to represent the local music scene. It is perhaps no surprise that both *Mad Love* and *Georgia* were very modest financial and popular successes, embroiled as they were in a milieu that was, by their respective release dates, somewhat passé.

CHAPTER 6

1   Bell and Valentine 1997: 155.
2   Zukin 1995: 2–3.
3   Michael D. Smith, 'The Empire Filters Back: Consumption, Production and the Politics of Starbucks Coffee', *Urban Geography*, 17, 6, 1996: 503.
4   Hall 1982: 304.
5   Zukin 1995: 3.
6   Bell and Valentine 1997: 190.
7   Smith 1996: 512.
8   Smith cites E. Galeano, *Open Veins of Latin America: Five Centuries of the Pillage of a Continent* (New York: Monthly Review Press, 1973).
9   Margaret Visser, *The Rituals of Dinner: The Origins, Evolution, Eccentricities and Meaning of Table Manners* (Harmondsworth: Penguin, 1993), 123.
10 Alan Beardsworth and Teresa Keil, *Sociology on the Menu* (London: Routledge, 1997), 105.
11 Beardsworth and Keil 1997: 110.
12 Paul Andrews, 'Java Jive: So Many Cafes, So Little Time. Espresso Havens Provide Reassuring Refuge', *Seattle Times*, 22 November 1987, Pacific Section, 14.
13 Ibid.
14 Smith 1996: 512.
15 Pierre Bourdieu, *Distinction: A Social Critique of the Judgement of Taste*, trans. Richard Nice (Cambridge, MA: Harvard University Press, 1984). In short, what was occurring was a process of transforming coffee from a mundane, quotidian product to one that could sustain the same process of discernment as a fine wine – to quote Starbucks' Dave Olson, "we didn't invent coffee. We just started paying it some attention" (Details from Terry McDermott, 'Cash

Crop', *Seattle Times*, 28 November 1993, Pacific Section: 8.) As McDermott writes, discussing the ascendancy of Starbucks, "wine people politely discuss favorite vintages … coffee people argue roasting methods with religious fervour", a debate dependant upon the mutual acquisition of what Bourdieu calls "cultural competence" (McDermott 1993: 8; Bourdieu 1984: 13).

16 Starbucks was founded by the Seattle residents Gordon Bowker, Zev Siegl and Jerry Baldwin. Technically speaking, Starbucks was not the first gourmet coffee seller. Bowker, Siegl and Baldwin aligned themselves with a trend from Northern California in the late 1960s, concerned with "high-quality, locally produced foods" and which "emphasised quality, freshness and handcrafting". The store imported its coffee beans from Peet's Coffee in Berkeley, California, one of only two speciality coffee roasters on the West Coast at that time. Details from Terry McDermott, 'Cash Crop', *Seattle Times*, 28 November 1993, Pacific Section: 8.

17 As Roger Sale argues, the original vote to establish Pike Place Market "as a historical district" reflected the influence of new urban activists in Seattle, who "wanted a different kind of city, a Seattle more urban and more urbane". Sale 1994a: 224–5.

18 Peter Williams and Neil Smith, 'From "Renaissance" to "Restructuring"', in Smith and Williams 1996: 6–7.

19 Erica Carter, James Donald and Judith Squires (eds) *Space and Place: Theories of Identity and Location* (London: Lawrence and Wishart, 1993), ix.

20 As widely noted, Pike Place Market remains a tourist 'must see' in Seattle, with an average of over nine million visitors per year in the late 1990s. Details from Naomi Dillon, 'Seattle's a nice place to visit, and hordes are on their way', *Seattle Times*, 28 May 1999, Business section.

21 Pierre Bourdieu, *Distinction: A Social Critique of the Judgement of Taste*, trans. Richard Nice (Cambridge, MA: Harvard University Press, 1984), 1. A recent example of this was the struggles on the part of champagne producers to ensure that champagne could only be identified with wine products made in the Champagne region of France.

22 Schultz was the former marketing director of Starbucks. At the time of the takeover, he was the owner of Il Giornale Coffee Company, a chain of four Seattle and Vancouver BC espresso shops.

23 Robin Updike, 'Brewing up a marketing plan: purchase of Starbucks was Howard Schultz's idea of a good blend', *Seattle Times*, 16 June 1987, Business section: C1.

24 Details from *Starbucks website*, "Timeline and History", 13 October 1999, www.starbucks. com.

25 Updike 1987: C1.

26 See, for example, Blake Fleetwood, 'The new elite and an urban renaissance', *New York Times Magazine*, 14 January 1979, or Penelope Lemov, 'Celebrating the city', in *Builder*, 7, February 1984: 90–7.

27 Stacy Warren's term, as outlined in 'This Heaven Gives Me Migraines: The problems and promise of landscapes of leisure', in James Duncan and David Ley (eds), *Place/Culture/ Representation* (London: Routledge, 1993), 173–86; Smith 1996a: 508.

28 Smith 1996a; Davis 1992.

29 Smith 1996a.

30 M. Christine Boyer, 'Cities for Sale: Merchandising History at South Street Seaport', in Michael Sorkin (ed.), *Variations on a Theme Park* (New York: Hill and Wang, 1992), 191–2.

31 *Cribbs Causeway homepage*, 12 October 1999, www.cribbs-causeway.co.uk.

32 In a profile for *The Sunday Times* magazine, Svenson stated that "my mission was to set up a genuine Seattle coffee bar where I could get the kind of coffee I missed" (12 July 1998: 35).

33 'Starbucks to roast Europe', *CNN fn*, www.cnnfn.com. 29 April 1998.

34 Phrase on Seattle Coffee Company pamphlet, 1999.

35 Seattle Coffee Company pamphlet, 1999.

36 Bell and Valentine 1997: 149.

37 Interestingly, such promotional activity by the Seattle Coffee Company also worked to echo the recent spatial and ideological reshaping of Seattle's urban landscape. In an article entitled 'Postindustrial Park or Bourgeois Playground?' (in Bennett and Teague (eds) 1990: 111–34), Richard Heyman examined the forces at work in the preservation and restructuring of Seattle's Gas Works Park. The park, which opened in 1975, was built on the site of the Seattle Gas Light Company's Lake Station Gassification Plant, and was "the world's first industrial site-conversion park". Heyman argued that the park, which incorporated "pieces of preserved industrial machinery within open green space", was predicated upon an "aestheticisation of the gas works" producing an "obsolescence narrative" that understood industrial equipment such as the outmoded gas works as part of Seattle's industrial past. Moreover, the park was constructed so as to draw users "into the city experience … look[ing] inward and embrac[ing] an aestheticised city figured in the burgeoning skyline of office buildings".

38 Nigel Thrift, 'Doing regional geography in a global system: the new international finance system, the City of London, and the south east of England, 1984–7', in R. Johnston, J. Hauer and G. Hoekveld (eds), *Regional Geography: Current Developments and Future Prospects* (London Routledge, 1990), 180.

39 Jessop Sutton, 'The Seattle Coffee Shop in Constantia Village', 14 June 1999, www.humanbeams.com/lavie/travel/constantia_coffee.shtml, and Starbucks promotional literature, 'Timeline', www.starbucks.com/company/timeline.

40 By the end of the 1990s, Starbucks had bequeathed urban and ex-urban centres across the globe with a whole range of new companies created in order to acquire a market share of the lucrative gourmet coffee business. Some, such as the UK chain Coffee Republic, or the Canadian retailer Second Cup, had no connection to Seattle in a proprietary sense. However, Starbucks, like Nike, McDonalds, or Microsoft, can be said to have invested their industry. In this sense, the connection of retailers such as Second Cup and Coffee Republic to Starbucks, and thus to the mobilising of images of Seattle-styled 'coffee culture' is implicit: without the discourse of gourmet coffee promulgated by Starbucks, it is arguable that they would not exist in the form that they do.

41 Details from the Business Wire, 18 November 1999, www.businesswire.com/webbox/bw.111899/193221527.

42 For example, coffee blends such as 'Yukon' trade on the companies Pacific Northwest heritage. Other details from Starbucks Coffee UK Booklet, 1999, and from the Starbuck's website.

43 Smith 1996a: 507.

44 Smith 1996a: 504.

45 For example, a *Time* magazine article on the city in November 1989, entitled 'Californians Keep Out!' made no mention of coffee in a quite lengthy discussion of Seattle's desirable attributes.

46 Mary T. Schmich, 'Now that its secret is out, Seattle is paying the price', *Chicago Tribune*, 29

March 1990.

47 Mary Bruno, 'Seattle Under Seige', *Lear's*, July 1991: 53.

48 Ann Japenga, 'On a Northwest Course', *Los Angeles Times*, 24 December 1992: E1, E8.

49 Richie Unterberger, *Seattle: Mini Rough Guide* (London: Rough Guides, 1998); *Lonely Planet online*, 15 June 2000, www.lonelyplanet.com/dest/nam/sea.htm; *Fodors.com MiniGuides*, www.fodors.com/miniguides/features/seattle_feat_0; Bruce Barcott, 'Coffee and Culture', in John Wilcock (ed.), *Insight Guides: Seattle*, (Singapore: APA Publications, 1997), 65.

50 Perhaps not surprisingly, the City of Seattle has not been averse to jumping on the coffee cart. The city's official website featured a 'virtual tour of Seattle' with text proclaiming that "Seattle is also known as the birthplace of the recent crazes for grunge rock and espresso coffee." (Seattle Virtual Tour, 14 June 2000, www.ci.seattle.wa.us/tour/intro.htm?331,133.)

51 Quoted in Chuck Taylor, 'Seattle's a bit player', *Seattle Times*, 27 April 1997, features section.

52 Steve Neale and Frank Krutnick, 'Broadcast Comedy and the Sitcom', in *Popular Film and Television Comedy* (London: Routledge, 1990), 222, 239.

53 Naomi Klein, *No Logo* (London: Flamingo, 2000), 44.

54 Klein 2000: 20.

55 *Fight Club*, dir. David Fincher, perf. Edward Norton, Brad Pitt and Helena Bonham Carter, Twentieth Century Fox, 1999.

56 Planet Hollywood is the permanent name for the chain of 'Hollywood'-themed restaurants, 'owned' by Arnold Schwarzenegger, Bruce Willis and Sylvester Stallone, and with interiors adorned by authentic movie memorabilia. The first outlet opened in New York City in 1984, and by the beginning of 2000, there were over seventy restaurants, stretching from Seattle to Singapore, and including outlets in Beirut and Tel Aviv (details from Planet Hollywood website, www.planethollywood.com). Planet Reebok was the title and the theme of an international advertising campaign by the sport shoe and apparel manufacturer in the late 1990s, with television clips directed by action cinematographer Bobby Carmichael. It was also the URL of a Reebok website, designed to coincide with the campaign, and subsequently cited frequently as an early example of successful internet marketing of a brand image (see, for example, the kyberco marketing pages at www.kyberco.com).

57 *In Like Flint* (1967).

58 Klein 2000: 136.

59 Ibid.

60 Mary Scott, 'An Interview with Howard Schultz, CEO of Starbucks Coffee Co.', *Business Ethics Magazine*, November/December 1995, www.depaul.edu/ethics/bizinter.html.

61 Scott 1995.

62 Bell and Valentine 1997: 195.

63 Bell and Valentine 1997: 194.

64 Bell and Valentine 1997: 199.

65 Bell and Valentine 1997: 196.

66 Stephanie Tate, 'Starbucks Agrees to Code of Conduct', *Stranger*, 15 March 1995: 9.

67 Ibid.

68 Klein 2000: 360–1.

69 Starbucks website, www.starbucks.com/community. The site includes information on 'The Starbucks foundation' a "non profit foundation [that] expands the historical connection between coffee and books by focusing its efforts on the cause of literacy", on Starbucks

community programmes, and Starbucks's environmental awareness, on issues such as recycling and organic produce.

70 Jamie Doward, 'And the brands play on…', *Observer*, 24 October 1999, Business section: 7.
71 Tomlinson 1999: 2.
72 Hattam 1995: 12.
73 Howard Schultz, quoted in 'Starbucks to Roast Europe', 1998.
74 Bell and Valentine 1997: 192.
75 Smith 1996a: 516.
76 Starbucks Coffee UK Booklet, 1999.
77 Smith 1996a: 517. Smith argues, convincingly, that contemporary consumption "in which the appeal of many of the most sought after commodities lies precisely in the putative authenticity derived from their origin in Third World production" makes Jameson's claim "clearly open to question" (503).
78 Smith 1996a: 521.
79 Giles Whittell, 'The language of hyper-choice has reached its frothy zenith in the coffee culture of America', *The Times*, 20 February 1999, Magazine section: 12.
80 *United States Trade Representative* homepage, 25 September 1999, www.ustrgov.
81 Figures from Michael Elliott, 'The New Radicals', *Newsweek*, 13 December 1999: 26–9, and from Patrick McMahon and James Cox, 'Seattle Clamps Down', *USA Today*, 1 December 1999: 1A; Kenneth Klee, 'The Siege of Seattle', *Newsweek*, 13 December 1999: 20–5.
82 Elliott 1999: 28
83 Klee 1999: 24.
84 Ibid.
85 Todd Gitlin, 'From Chicago to Seattle', *Newsweek*, 13 December 1999: 2.
86 Klee 1999: 22–3.
87 The attack of Starbucks was also used by the American press to highlight what they saw as the paradox of the demonstrations against the WTO talks. *Newsweek* carried a cartoon by Ed Stein, originally published in the *Denver Rocky Mountain News*, depicting two protestors taking a break from marching. Holding placards stating 'Free Trade Kills' and 'Down with the WTO', the protestors were shown leaning against the service counter of a Starbucks coffee store, and being asked by the barrister for their choice of "Costa Rican, Guatemalan, or Sumatra?" When combined with the company's rapid rise to ubiquity, and its allegedly aggressive real estate practices, Starbucks' deliberate emphasis of the relations of production and its 'coffee tourism' ensured that it would be a high-profile target, at least for the attention of the American media.
88 A number of downtown stores that suffered 'major damage' according to reports carried by *Newsweek*, and the *Seattle Times*, 1 December 1999.
89 Bell and Valentine 1997: 191.
90 Zukin 1995: 3.

CONCLUSION

1 Seattle Mayor Paul Schell, quoted in Josh Fiet, 'Boarding Up Boomtown', *Stranger*, 9, 12, 9 December 1999: 5.
2 Patrick McMahon and James Cox, 'Seattle Clamps Down', *USA Today*, 1 December 1999: 1A.

3   Sam Howe Verhovek, 'Seattle is Stung, Angry and Embarrassed as Opportunity Turns to Chaos', *New York Times*, 2 December 1999: 1.

4   Klee 1999: 20–5.

5   Andrew Gumbel, 'Trading Insults and Pepper Spray in a Global Row', *Independent*, 3 December 1999: 17.

6   Giles Whittell, 'Jobless in Seattle as slump bites', *The Times*, 5 December 1998: 18.

7   Ibid.

8   Beauregard 1993: 220; Kennedy 2000: 17.

9   Kevin Robins, 'Prisoners of the City: Whatever Could a Postmodern City Be?', in Erica Carter, James Donald and Judith Squires (eds), *Space and Place: Theories of Identity and Location* (London: Lawrence and Wishart, 1993), 303.

10 Robins 1993: 312.

11 Robins 1993: 325.

12 Kevin Robins, 'Tradition and translation: national culture in its global context', in Jessica Evans and David Boswell (eds), *Representing the Nation, A Reader: Histories, Heritage and Museums* (London: Routledge, 1999), 35.

BIBLIOGRAPHY

Abbott, Carl (1993) *The Metropolitan Frontier*. Tucson: University of Arizona Press.
Abcarian, Robin (1991) 'Boomer Backlash', *Los Angeles Times*, 12 July, Section E1.
Adler, Jerry (1996) 'Seattle Reigns', *Newsweek*, 20 May, 48–59.
Aglietta, Michel (1979) *A Theory of Capitalist Regulation: The US Experience*. NLB: London.
Alexie, Sherman (1998) *Indian Killer*. London: Vintage.
___ (1993) *The Lone Ranger and Tonto Fistfight in Heaven*. New York: The Atlantic Monthly Press.
___ (1996) *Reservation Blues*. London: Minerva.
Anon. (1975) 'Ranking the Cities', *Time*, 29 September, 39–40.
___ (1977) 'Dixy Rocks the Northwest', *Time*, 12 December, 26–36.
___ (1989a) 'Don't Block the View', *The Economist*, 27 May, 30.
___ (1989b) 'Seattle's Welcome Mat Begins to Fray', *US News and World Report*, 6 November, 16.
___ (1990) 'From Fishmonger to Pastrami Stand?', *The Economist*, 10 February, 25.
___ (1993a) 'Forget the "X" Please', *Mediaweek*, 2 August, 10.
___ (1993b) "Marketing Tools", *American Demographics*, Fall, 1.
___ (1995a) 'Seattle Commons', *Progressive Architecture*, January, 102–3.
___ (1995b) 'Seattle: towards a democratic Nirvana?', *New Statesman and Society*, 24 March, 29–30.
___ (1998a) 'Ally Svenson: Coffee Merchant', *The Sunday Times Magazine*, 12 July, 35.
___ (1998b) 'Starbucks to roast Europe', *CNN*. 29 April. www.cnnfn.com.
___ (1999) 'NAACP Blasts TV Networks' Fall Season Whitewash', *NAACP Homepage*. 12 July. www.naacp.org/president/releases/naacp_blasts_tv_networks.htm.
___ (2000) 'Emerging Markets for Outdoor Recreation in the United States', *ORCA website*. 12 June. www.outdoorlink.com/infosource/nsre/chap2/CH2.htm.
Amin, Ash (ed.) (1994) *Post-Fordism*. Oxford and Cambridge, MA: Basil Blackwell, 1994.

Anderson, Benedict (1993) *Imagined Communities*. London: Verso.

Anderson, Kay and Fay Gale (eds) (1992) *Inventing Places: Studies in Cultural Geography*. Melbourne: Longman Cheshire.

Anderson, Rick (1988) 'Cruising our livable city's new killing grounds', *Seattle Times*, 28 November, B1

Anderson, Ross (1987) 'Putting "Best" To The Test', *Seattle Times*, 15 February. Pacific Section, 9.

Andrews, Paul (1987) 'Java Jive: So Many Cafes, So Little Time. Espresso Havens Provide Reassuring Refuge', *Seattle Times*, 22 November, Pacific Section, 14.

Appadurai, Arjun (1990) 'Disjuncture and Difference in the Global Cultural Economy', in *Theory, Culture and Society*, 7, 295–310.

Ashworth, G. J. and H. Voogd (1990) *Selling The City*. London: Belhaven Press.

Bagge, Peter (1994) *Buddy the Dreamer*. Seattle: Fantagraphics Books.

___ (1998) *Hey Buddy!* Seattle: Fantagraphics Books.

Bakhtin, Mikhail (1981) *The Dialogic Imagination*, ed. Michael Holquist, trans. Caryl Emerson and Michael Holquist. Austin: University of Texas Press.

Baker, H. (1985) 'Streetwise: A True Tale of the City', *American Cinematographer*, LXVI/9, September, 44–8.

Barcott, Bruce (1997) 'Coffee and Culture', *Insight Guides: Seattle*, ed. John Wilcock. Singapore: APA Publications, 65–70.

Barcus, Neil W. (1996) 'Seattle's Jimi', *Stranger*, 12 June.

Barker, Martin and Kate Brooks (1998) *Knowing Audiences: Judge Dread*. Luton: University of Luton Press.

Beardsworth, Alan and Teresa Keil (1997) *Sociology on the Menu*. London: Routledge.

Beauregard, Robert A. (1993) *Voices of Decline*. Cambridge, MA: Blackwell.

Becker, Elizabeth (1992) 'Private Idaho', *New Republic*, 4 May, 9–10.

Bell, Daniel (1973) *The Coming of Post-industrial Society*. New York: Basic Books.

Bell, David and Gill Valentine (1997) *Consuming Geographies*. London: Routledge.

Bellafante, Gina (1993) 'Where's the Next Seattle?', *Time*, 25 October, 66.

Bellah, Robert N. (1985) *Habits of the Heart*. San Francisco: Harper and Row.

Benjamin, Walter (1983) *Charles Baudelaire: A Lyric Poet in the Era of High Capitalism*, trans. Harry Zohn. London: Verso.

Bennett, Michael and David W. Teague (eds) (1999) *The Nature of Cities: Ecocriticism and Urban Environments*. Tucson: University of Arizona Press.

Bennett, Tony and Janet Woollacott (1987) *Bond and Beyond: The Political Career of a Popular Hero*. London: MacMillan Education.

Bennet, Tony, Colin Mercer and Janet Woollacott (eds) (1986) *Popular Culture and Social Relations*. Milton Keynes: Open University Press.

Berman, Marshall (1983) *All That Is Solid Melts Into Air*. London: Verso.

Bhabha, Homi (1990) 'Novel Metropolis', *New Statesman and Society*, 16 February.

Bird, Jon, Barry Curtis, Tim Putnam, George Robertson and Lisa Tickner (eds) (1993) *Mapping the Futures*. London: Routledge.

Blumin, Stuart M. (1984) 'Explaining the New Metropolis', *Journal of Urban History*, 11, 1, November, 9–38.

Bock, Audie (1984) 'Local Heroes', *American Film*, 4, 6, 38–42.

Bocock, Robert (1993) *Consumption*. London: Routledge.

Bocock, Robert, Robert Thompson and Kenneth Thompson (eds) (1992) *Social and Cultural Forms of Modernity*. Cambridge: Polity Press.

Bonfante, Jordan (1989) 'Californians Keep Out!', *Time*, 13 November, 38–9.

Bourdieu, Pierre (1984) *Distinction: A Social Critique of the Judgement of Taste*, trans. Richard Nice. Cambridge, MA: Harvard University Press.

Bowermaster, Jon (1991) 'Seattle: Too Much of a Good Thing?', *New York Times Magazine*, 6 January, 24–42.

Boyer, M. Christine (1986) *Dreaming the Rational City*. London: MIT Press.

___ (1992) 'Cities for Sale: Merchandising History at South Street Seaport', in Michael Sorkin (ed.) *Variations on a Theme Park*. New York: Hill and Wang, 191–2.

___ (1993) 'The City as Illusion: New York's public places', in Paul L. Knox (ed.) *The Restless Urban Landscape*. New Jersey: Prentice Hall.

___ (1994) *The City of Collective Memory*. Cambridge, MA: MIT Press.

Brooker, Peter (1996) *New York Fictions: Modernity, Postmodernism, The New Modern*. New York: Longman.

Brophy, Beth (1995) 'Stressless – and Simple – in Seattle', *US News and World Report*, 11 December, 96–7.

Brotchie, J., Mike Batty, Ed Blakely, Peter Hall and Peter Newton (eds) (1995) *Cities in Competition*. Melbourne: Longman.

Bruno, Giuliana (1987) 'Ramble city: postmodernism and *Blade Runner*', *October*, 41: 61–74.

Bruno, Mary (1991) 'Seattle Under Siege', *Lear's*, July, 53.

Buchanan, R. A. (1965) *Technology and Social Progress*. London: Pergamon Press.

Burkeman, Oliver and Emma Brockes (1999) 'Trouble Brewing', *Guardian*, 3 December, G2, 2–3.

Business Wire (1999) 12 December, www.businesswire.com/webbox/bw.111899/193221527.

Canty, Donald (1992) 'Seattle Affordability', *Progressive Architecture*, May, 2–6.

___ (1993) 'Seattle Conference on Reshaping Cities', *Progressive Architecture*, November, 25.

Carlin, John (1996) 'Seattle's first black mayor cleans up a white city', *Independent*, 15 July.

Carter, Erica, James Donald and Judith Squires (eds) (1993) *Space and Place: Theories of Identity and Location*. London: Lawrence and Wishart.

Case, Frederick (1989) 'How Livable? Looking at No. 1: Seattle annoyed to receive the top rating as America's most livable city', *Seattle Times*, 27 October, E1.

Castells, Manuel (1989) *The Informational City*. Oxford: Blackwell.

Chafe, William (1998) *The Unfinished Journey: America Since World War Two*. Oxford: Oxford University Press.

Chaney, David (1997) *Lifestyles*. London: Routledge.

Church, George J. (1989) 'Urban Growing Pains', *Time*, 29 May, 33.

City of Seattle Official Website (2000) www.ci.seattle.wa.us/tda/dstour.htm.

Clarke, David B. (ed.) (1997) *The Cinematic City*. London: Routledge.

Coddington, Grace (1992) 'Grunge and Glory', *Vogue*, December, 254–63.

Coleman, Robin R. Means (1998) *African American Viewers and the Black Situation Comedy*. New York: Garland.

Collins, Jim, Hilary Radner and Ava Preacher Gardner (eds) (1993) *Film Theory Goes to the Movies: Cultural Analysis of Contemporary Film*. New York: Routledge.

Committee for the Seattle Commons (1995) 'Seattle Commons', *Progressive Architecture*, January, 102–3.

Connelly, Joel (1993) 'Border Patrol U.S., Canada keep a concerned eye on suburban sprawl', *Chicago Tribune*, 14 November.

Coontz, Stephanie (1992) *The Way We Never Were*. New York: Basic Books.

Cosgrove, Denis E. (1984) *Social Formation and Symbolic Landscape*. London: Croom Helm.

Coupland, Douglas (1991) *Generation X*. New York: St Martin's Press.

___ (1995) *Microserfs*. London: Flamingo.

Crang, Mike (1998) *Cultural Geography*. London: Routledge.

Cribbs Causeway (1999) 14 November. www.cribbs-causeway.co.uk.

Crichton, Michael (1994) *Disclosure*. London: Arrow.

Cringely, Robert X. (1996) *Accidental Empires*. London: Penguin.

Crowe, Cameron (1993) 'Five against the world', *Rolling Stone*, 28 October, 56.

Daniels, Roger (1988) *Asian America*. Seattle: University of Washington Press.

Davies, Jude '"I'm the Bad Guy?": *Falling Down* and White Masculinity in 1990s Hollywood', *Journal of Gender Studies*, 4, 2, 145–52.

Davies, Sophie (1999) 'Speechless in Seattle', *She*, September, 196.

Davis, Mike (1990) *City of Quartz*. London: Vintage.

___ (1999) *Ecology of Fear*. London: Picador.

De Certeau, Michel (1988) *The Practice of Everyday Life*. Berkeley: University of California Press.

Dear, Michael R. (2000) *The Postmodern Urban Condition*. Oxford: Blackwell.

De Leon, Ferdinand M. (1996) 'City's charms continue to lure visitors to area', *Seattle Times*, 16 August.

Denzin, Norman K. (1991) *Images of Postmodern Society*. London: SAGE.

Dillon, Naomi (1991) 'Seattle's a nice place to visit, and hordes are on their way', *Seattle Times*, 28 May, Business Section.

DiMartino, Dave (1990) 'A Seattle Slew', *Rolling Stone*, 20 September, 23.

Donald, James (1992) 'Metropolis: The City as Text', in Robert Bocock, Robert Thompson and Kenneth Thompson (eds), *Social and Cultural Forms of Modernity*. Cambridge: Polity Press, 418–59.

___ (1997) 'This, Here, Now: Imagining the Modern City', in Sallie Westwood and John Williams (eds) *Imagining Cities*. London: Routledge, 181–201.

Donaton, Scott (1993) 'The Media Wakes Up To Generation X', *Advertising Age*, 64, 5, 1 February.

Doward, Jamie (1999) 'And the brands play on…', *Observer*, 24 October, Business section, 7.

Duff, Christina (1998) 'It's Sad but True: Good Times are Bad for Real Slackers', *Wall Street Journal*, 6 August, A1–A5.

Duhigg, Charles (1999) 'Means of Dissent', *New Republic*, 20 December, 14.

Dunant, Sarah and Roy Porter (eds) (1996) *The Age of Anxiety*. London: Virago.

Duncan, James and David Ley (eds) (1993) *Place/Culture/Representation*. London: Routledge.

DuttaAhmed, Shantanu (1998) 'Heartbreak Hotel: MTV's *The Real World, III*, and The Narratives of Containment', *American Studies*, 39, 2, 157–71.

Dyer, Richard (1988) 'White', *Screen*, 28, 4, 44–64.

___ (1997) *White*. London: Routledge.

Eagleton, Terry (1991) *Ideology*. London: Verso.

Eddie Bauer online (1997) 'Seattle's Best Coffee and Eddie Bauer Unveil Proprietary Coffee Blend', 27 October. www.eddiebauer.com/about/frame_971027.asp.

___ (2000) 'Company background', 5 May. www.eddiebauer.com.

Egan, Timothy (1988) 'REI: Three Initials That Changed Life In The Northwest', *Seattle Times*, 6 March, Pacific Section, 4.

___ (1989) *Portrait of Seattle*. Portland, Oregon: Graphic Arts Center Publishing.

Ehrenreich, Barbara (1989) *Fear of Falling*. New York: Pantheon Books.

Ehrenreich, Barbara and John Ehrenreich (1979) 'The professional-managerial class', in Pat Walker (ed.) *Between Labour and Capital*. Boston: South End Press.

Elkins, Aaron (1993) *Old Scores*. New York: Ballantine Books.

Elliott, Michael (1999) 'The New Radicals', *Newsweek*, 13 December, 26–9.

Emerson, Earl (1988) *Black Hearts and Slow Dancing*. New York: Avon Books.

___ (1996) *The Vanishing Smile*. New York: Ballantine Books.

Epstein, Jonathon S. (ed.) (1999) *Youth Culture: Identity in a Postmodern World*. Oxford: Blackwell.

Ericsson, Celeste (1990) 'California Bashing – This isn't lively debate; it's ugliness', *Seattle Times*, 21 October, Issues Section, A23.

Faircloth, Anne and Geoffrey Precourt (1996) 'Best Cities: Where the Living is Easy, *Fortune*, 11 November, 102–6.

Farley, Christopher John (1993) 'Rock's Anxious Rebels', *Time*, 25 October, 60.

Fefer, Mark D. (1997) 'Is Seattle the next Silicon Valley?', *Fortune*. 7 July. www.pathfinder.com/@@Btj*4AUAotoeCma9/1997/970707/sea.html.

Ferigno, Robert (1997) 'Weatherproof in Seattle', *New York Times Magazine*, 11 May, 53–82.

Fey, William H. (1993) 'The New Urban Revival in the United States', *Urban Studies*, 30, 4/5, 741–74.

Fiet, Josh (1999) 'Boarding Up Boomtown', *Stranger*, 9, 12, 9 December, 5.

Findlay, John M. (1992) *Magic Lands*. Berkeley and Los Angeles: University of California Press.

___ (1997) 'A Fishy Proposition: Regional Identity in the Pacific Northwest', in Michael C. Steiner and David M. Wrobel (eds) *Many Wests*. Lawrence, KA: University Press of Kansas, 37–70.

Fiske, John (1989) *Television Culture*. London: Routledge.

Flores, Michael Matassa (1999) 'Seattle left less naïve as it counts physical, psychological costs', *Seattle Times*, 5 December, 1.

Fodors.com (2000) 'MiniGuides', 2 July. www.fodors.com/miniguides/features/seattle_feat_01.

Ford, G. M. (1996) *Cast in Stone*. New York: Avon Books.

Foucault, Michel (1979) *Discipline and Punish*. London: Penguin.

___ (1986) 'Of Other Spaces', *Diacritics*, Spring, 22–7.

Francke, Lizzie (1992) '*American Heart*', *Sight and Sound*, May, 40–1.

Frankenberg, Ruth (1993) *White Women, Race Matters*. Minneapolis: University of Minnesota Press.

Frascella Lawrence (1999) 'Nora Ephron', *Rolling Stone*, 8 July, 73–5.

Fresh Cup (1999) 27 September. www.freshcup.com/mediakit/accolades.html.

Frey, William H. (1992) 'Boomer Magnets', *American Demographics*, March, 34–53.

Frutig, Judith (1975) 'America's Most Livable City', *Christian Science Monitor*, 21 May, 27–8.

Gabriel, John (1998) *Whitewash: Racialised Politics and the Media*. London: Routledge.

Gaines, Donna (1991) *Teenage Wasteland*. New York: Pantheon Books.

Galeano, E. (1973) *Open Veins of Latin America: Five Centuries of the Pillage of a Continent*. New

York: Monthly Review Press.

Garreau, Joel (1991) *Edge City*. New York: Doubleday.

Gates, Bill (1996) *The Road Ahead*. London: Penguin.

Geddes, Robert (ed.) (1997) *Cities in Our Future*. Washington DC: Island Press.

Gehman, Pleasant (1992) 'Artist of the Year: Nirvana', *Spin*, December, 51

Gelder, Ken and Sarah Thornton (1997) *The Subcultures Reader*. London: Routledge.

Gerbner, George (2000) 'Casting the American Scene: Fairness and Diversity in Television', *Screen Actors Guild website*, 12 May. www.sag.com/special/americanscene.html.

Gilmore, Susan and Helen Jung (1999) 'WTO organizers "glad we did this"', *Seattle Times*, 4 December.

Gitlin, Todd (1999) 'From Chicago to Seattle', *Newsweek*, 13 December, 2.

Gordon, Michael (ed.) (1978) *The American Family in Social-Historical Perspective*. New York: St. Martin's Press.

Gottlieb, Joanne and Gayle Wald (1994) 'Smells Like Teen Spirit: Riot Grrrls, Revolution and Women in Independent Rock', in Andrew Ross and Tricia Rose (eds) *Microphone Friends: Youth Music and Youth Culture*. London: Routledge, 250–74.

Gramsci, Antonio (1973) *Selections from the Prison Notebooks of Antonio Gramsci*, eds Quintin Hoare and Geoffrey Nowell-Smith. London: Lawrence and Wishart.

Gray, Herman (1995) *Watching Race: Television and the Struggle for Blackness*. Minneapolis: University of Minneapolis Press.

The Greater Seattle Datasheet 1999–2000. 4 July 2000. www.ci.seattle.wa.us/oir/facts.htm

Griffith, Thomas (1976) 'The Pacific Northwest', *Atlantic Monthly*, April, 46–93.

Gross, David M. and Sophfronia Scott (1990) 'Proceeding With Caution', *Time*, 16 July, 57.

Grossberg, Lawrence (1997a) *Bringing It All Back Home*. North Carolina: Duke University Press.

___ (1997b) *Dancing in Spite of Myself*. North Carolina: Duke University Press.

Gubar, Susan (1997) *Racechanges: White Skin, Black Face in American Culture*. Oxford: Oxford University Press.

Guerrero, Ed (1993) *Framing Blackness: The African American Image in Film*. Philadelphia: Temple University Press.

Gumbel, Andrew (1998) 'Taking a walk on the wild side … in Seattle', *Independent on Sunday*, 20 September, 16.

___ (1999) 'Trading Insults and Pepper Spray in a Global Row', *Independent*, 3 December, 17.

Guterson, David (1995) *Snow Falling on Cedars*. London: Bloomsbury.

Hall, Stuart (1982) 'The Rediscovery of "Ideology": The Return of the "Repressed" in Media Studies', in Michael Gurevitch, Tony Bennett, James Curran and Janet Wollacott (eds) *Culture, Society and the Media*. London: Methuen, 56–90.

___ (1992a) 'What is this "Black" in Black Popular Culture?', in Gina Dent (ed.) *Black Popular Culture, a Project by Michelle Wallace*. Seattle: Bay Press, 21–33.

___ (1992b) 'The Question of Cultural Identity', in Stuart Hall, David Held and Tony McGrew (eds) *Modernity and Its Futures*. Cambridge: Polity Press, 273–325.

Hall, Tim and Phil Hubbard (1996) 'The entrepreneurial city: new urban politics, new urban geographies?', *Progress in Human Geography*, 20, 2, 153–75.

Hamamoto, Darrell Y. (1994) *Monitored Peril: Asian Americans and the Politics of TV Representation*. Minneapolis: University of Minnesota Press.

Hamel, Ruth and Tim Schreiner (1989) 'The Bad Side of Good Times', *American Demographics*,

February, 46–8.

Harvey, David (1989) *The Condition of Postmodernity*. Oxford: Blackwell.

___ (1996) *Justice, Nature and the Geography of Difference.* Oxford: Blackwell.

Hattam, Jennifer (1995) 'Grounds For Dispute: Activists Target Starbucks', *Stranger*, 3 January, 12.

Hendricksen, Louise (1995) *Lethal Legacy*. New York: Zebra Books.

Herting, Jerald R., David B. Grusky and Steven E. Van Rompaey (1997) 'The Social Geography of Interstate Mobility and Persistence', *American Sociological Review*, 62, 2.

Heyman, Richard (1999) 'Postindustrial Park or Bourgeois Playground?: Preservation and Urban Restructuring at Seattle's Gas Works Park', in Michael Bennett and David W. Teague (eds), *The Nature of Cities: Ecocriticism and Urban Environments.* Tucson: University of Arizona Press, 111–34.

Hinterberger, John (1989) 'The "C" Word: Guess what state she's from?', *Seattle Times*, 2 August, Scene Section, E1.

Hodgins, Randy and Steve McLellan (1995) *Seattle on Film*. Olympia: Punchline Productions.

Hodgkinson, Tom (1995) 'Age of the cool nerd', *Guardian*, 23 October, 10–11.

Hoffelt, Mary (1976) 'Coffee: 300 Million Cups a Day', *Seattle Post-Intelligence*, 25 January, Northwest Section, 4–11.

hooks, bell (1992) *Black Looks*. Boston: South End Press.

Houston, Eric (1994) 'Seattle: The Nation's "Gateway to Asia"', *Black Enterprise*. May, 62–3.

Howe, Neil and William Strauss (1992) 'The New Generation Gap', *Atlantic Monthly*, December, 67–89.

Hoyt, Richard (1981) *30 For a Harry*. New York: Penguin.

___ (1985) *Fish Story*. New York: Tom Doherty Associates.

Humphrey, Clark (1995) *Loser: The Real Seattle Music Story*. Portland: Feral House.

Hunter, Jennifer (1997) 'The Deadly Streets', *Maclean's*, 8 December, 32–3.

Iglitzin, Lynne B. (1995) 'The Seattle Commons: A Case Study in the Politics and Planning of an Urban Village', *Policy Studies Journal*, 23, 4, 620–35.

Imrie, Rob, Steven Pinch and Mark Boyle (1996) 'Identities, Citizenship and Power in Cities', *Urban Studies*, 33, 1255–61.

Jacklett, Ben (1999) 'Truncated Monorail?', *Stranger*, 8, 17, 14 January.

Jackson, John Brinckerhoff (1994) *A Sense of Place, a Sense of Time*. New York: Yale University Press.

Jacobs, Jane (1962) *The Death and Life of Great American Cities*. London: Jonathan Cape.

Jakle, John A. (1985) *The Tourist: Travel in Twentieth-Century North America.* Lincoln: University of Nebraska Press.

Jameson, Fredric (1991) *Postmodernism, or the Cultural Logic of Late Capitalism*. London: Verso.

Jance, J. A. (1991) *Payment in Kind*. New York: Avon Mystery.

___ (1992) *Without Due Process*. New York: Avon Mystery.

Jancovich Mark and James Lyons (eds) (2003) *Quality Popular Television*. London: BFI.

Japenga, Ann (1993) 'Grunge 'R' Us', *Los Angeles Times Magazine*, 14 November, 26, 45.

Jarvis, Brian (1998) *Postmodern Cartographies*. London: Pluto Press.

Jenkins, Nancy Harmon (1992) 'As American as Apple Pie', *New York Times Magazine*, 1 March, 67.

Jewell, Mark (2000) 'Tourist industry makes strides in Washington', *Seattle Times*, 14 February.

Jewson, Nick and Susanne MacGregor (eds) (1997) *Transforming Cities*. London: Routledge.

Jones, Mary Ellen (ed.) (1994) *The American Frontier: Opposing Viewpoints*. San Diego: Greenhaven Press.

Judd, Ron C. (1997) 'Ron C. Judd's Trail Mix: Seeking to prove true NW heritage?', *Seattle Times*, Sports News, 16 January.

Kaplan, Amy and Donald E. Pease (eds) (1993) *Cultures of United States Imperialism*. London: Duke University Press.

Kasinitz, Philip (1995) *Metropolis: Centre and Symbol of Our Times*. London: MacMillan Press.

Kasson, John F. (1978) *Civilizing the Machine: Technology and Republican Values in America, 1776–1900*. New York: Grossman.

Kennedy, Liam, (1996) 'Alien Nation: White Male Paranoia and Imperial Culture in the United States', *Journal of American Studies*, 30, 1, 87–100.

___ (2000) *Race and Urban Space in Contemporary Culture*. Edinburgh: Edinburgh University Press.

King, Anthony D. (ed.) (1991) *Culture, Globalisation and the World System*. London: MacMillan.

Kitts, Jeff, Brad Tolinski and Harold Steinblatt (eds) (1998) *Nirvana and the Grunge Revolution*. Milwaukee: Hal Leonard.

Klee, Kenneth (1999) 'The Siege of Seattle', *Newsweek*, 13 December, 22–3.

Knox, Paul L. and Peter J. Taylor (eds) (1995) *World Cities in a World System*. Cambridge: Cambridge University Press.

Koolhaas, Rem (1978) *Delirious New York: A Retroactive Manifesto for Manhattan*. London: Thames and Hudson.

Kresl, Peter Karl and Gary Gappert (eds) (1995) *North American Cities and the Global Economy*. Thousand Oaks, CA: Sage.

Kristeva, Julia (1986) 'Word, Dialogue and Novel', trans. Alice Jardine, Thomas Gora and Leon S. Roudiez, in Toril Moi (ed.) *The Kristeva Reader*. New York: Columbia University Press, 34–61.

Kugiya, Hugo (1997) 'Who we are and what we are becoming', *Seattle Times*, 3 August 1997, Features Section, 2.

kyberco marketing online (1999) 10 March. www.kyberco.com.

Labich, Kenneth (1993) 'The Best Cities for Knowledge Workers', *Fortune*, 15 November, 50–6.

Lane, Polly (1999) 'New owner of Seattle Magazine promises new focus, "new aura"', *Seattle Times*, 20 April, 3.

Lefebvre, Henri (1974) *La Production de l'espace*. Paris: Anthropos.

Lim, Paul J. (1996) 'Step Aside Boston', *Seattle Times*, 21 April, 2.

Lipsky, David (1996) 'Junkie Town', *Rolling Stone*, 30 May, 36–62.

Louis, Arthur M. (1975) 'The Worst American City', *Harper's*, January, 67–104.

Lucas, Eric (1997) 'Plane Facts about Boeing', in John Wilcock (ed.) *Seattle*. Singapore, APA Publications, 26–8.

McArthur, Colin (1997) 'Chinese Boxes and Russian Dolls: Tracking the Elusive Cinematic City', in David B. Clarke (ed.) *The Cinematic City*. London: Routledge, 19–41.

McOmber, J. Martin (1999) 'Dutch architect selected to design new library', *Seattle Times*, 27 May.

MacDonald, J. Fred (1992) *Blacks and White TV*. Chicago: Nelson-Hall.

MacDonald, Norbert (1987) *Distant Neighbors*. Lincoln: University of Nebraska Press.

McDermott, Terry (1993) 'Cash Crop', *Seattle Times*, 28 November, Pacific Section, 8.

McMahon, Patrick and James Cox (1999) 'Seattle Clamps Down', *USA Today*, 1 December, 1A.

Mandel, Ernest (1975) *Late Capitalism*. London: Verso.

Mark, Mary Ellen and Cheryl McCall (1993) 'Streets of the Lost', *Life*, July 1983, 34–42.

Martin Jr., Waldo E. (1995) 'Soulville in Seattle: African Americans, The City and the Paradox of the American Dream', *Reviews in American History*, 23, 643–9.

Marx, Karl and Friedrich Engels (1974) *The German Ideology*. London: Lawrence and Wishart.

Marx, Leo (1964) *The Machine in the Garden*. Oxford: Oxford University Press.

Massey, Doreen (1995) 'The conceptualisation of place', in Doreen Massey and Pat Jess (eds) *A Place in the World?* Oxford: Open University Press, 45–85.

Mazzoleni, Donna (1993) 'The City and the Imaginary', in Erica Carter, James Donald and Judith Squires (eds) *Space and Place: Theories of Identity and Location*. London: Lawrence and Wishart, 285–302.

McLaughlin Green, Constance (1957) *American Cities in the Growth of the Nation*. London: The Athlone Press.

Medved, Michael (1994) 'Paradise Lost – Riots, Fire, Slump and now Earthquake – for many Californians the dream is over', *The Sunday Times*, 23 January.

Mintz, Steven and Susan Kellogg (1988) *Domestic Revolutions*. New York: The Free Press.

Mitchell, W. J. T.(1986) *Iconology*. Chicago: University of Chicago Press.

___ (1994) *Picture Theory*. Chicago: University of Chicago Press.

Moody, Fred (1996) *I Sing The Body Electronic*. New York: Coronet.

Morgan, Murray (1962) *Skid Road: An Informal Portrait of Seattle*. New York: Viking Press.

Morley, David and Kevin Robins (1995) *Spaces of Identity*. London: Routledge.

Morrell, Brad (1996) *Nirvana and The Sound of Seattle*. London: Omnibus Press.

Morrill, Richard (1995) 'Racial Segregation and Class in a Liberal Metropolis', *Geographical Analysis*, 27, 1, January, 22–41.

Moscio, Inga (1993) 'Mister Sub Pop: Jonathan Poneman', *Stranger*, 27 September, 3.

Mumford, Louis (1961) *The City in History: Its Origins, Its Transformations and Its Prospects*. New York: Harcourt, Brace and World.

Murakami, Kery (1993) 'Espresso Goes East: New Yorkers Wake Up and Smell the Coffee from Seattle', *Seattle Times*, 11 July, Business Section, D1.

Nelson, Robert T. (1986) 'One more organisation discovers Seattle can be a great place to live', *Seattle Times*, 13 November, Northwest Section, F4.

Olson, Jack (1991) *Predator: Rape, Madness and Injustice in Seattle*. New York: Delacorte Press.

Paddison, Ronan (1993) 'City Marketing, Image Reconstruction and Urban Regeneration', *Urban Studies*, 30, 2, 339–50.

Page, Clarence (1991) 'Lament of the twentysomethings', *Chicago Tribune*, 11 August, 3.

Page, Stephen (1995) *Urban Tourism*. London: Routledge.

Park, Robert E., Ernest W. Burgess and R. D. McKenzie (eds) (1984) *The City*. Chicago: University of Chicago Press.

Patton, Dean (1999) 'Seattle's rise as the capital of the New Economy', *Christian Science Monitor*, 30 November, 4.

Patrick, Colleen (1990) 'Enough Already! On Bashing Californians', *Seattle Times*, 28 October, A19.

Perret, Eric (1993) 'A Houseboat in Seattle', *Esquire*, February, 32.

Pfeil, Fred (1995) *White Guys*. London: Verso.

Pierce, J. Kingston (1998) 'What We Love About Seattle: From A to Z', *Seattle*, September, 39–43.

Pile, Steve, Christopher Brook and Gerry Mooney (1999) *Unruly Cities?*. London: Routledge.

Pileggi, Sarah (1982) 'Seattle: City Life at its Best', *Sports Illustrated*, 19 July, 54–68.

Pollock, Griselda (1988) *Vision and Difference: Femininity, Feminism and the Histories of Art*. London: Routledge.

Poneman, Jonathan (1992) 'Grunge and Glory', *Vogue*, December, 260

Posner, Ellen (1991) 'A City That Likes Itself', *The Atlantic Monthly*, July, 94–100.

Precourt, Geoffrey and Anne Faircloth (1996) 'Best Cities: Where the Living is Easy', *Fortune*, 11 November 1996.

Raban, Jonathan (1990) *Hunting Mr Heartbreak*. London: Collins Harvill.

___ (1998) *Soft City*. London: Harvill.

Reed, J. D. (1985) 'All Riled Up About Ratings', *Time*, 11 March, 46.

Reeves, Tracey A. (1998) 'Harsh lives take heavy toll on Native Americans' health', *Seattle Times*, 3 August, Health and Science Section.

Robertson, George (ed.) (1996) *Futurenatural*. London: Routledge.

Robins, Kevin (1996) *Into the Image*. London: Routledge.

___ (1999) 'Tradition and translation: national culture in its global context', in Jessica Evans and David Boswell (eds) *Representing the Nation: A Reader: Histories, Heritage and Museums*. London: Routledge, 15–32.

Roe, JoAnn (1989) 'Seattle: The Emerald City', *American West*, April, 42–9.

___ (1996) *Seattle Uncovered*. Texas: Seaside Press.

Rogers, Richard (1997) *Cities for a Small Planet*. London: Faber and Faber.

Rose, David C. (1990) 'City Profile: Seattle', *Cities*, 7, 4, 283–9.

Rosenberg, Matt (1999) 'Seattle Politics, Always a Riot', *Weekly Standard* Magazine, 5, 13, 13.

Ross, Andrew (1994) *The Chicago Gangster Theory of Life*. London: Verso.

Rothenberg, Randall (1995) 'What Makes Sammy Walk?' *Esquire*, May, 72–9.

Rule, Anne (1994) *The Stranger Beside Me*. London: Warner Books.

Rutheiser, Charles (1996) *Imagineering Atlanta*. New York: Verso.

Sale, Roger (1994a) *Seattle, Past to Present*. Seattle: University of Washington Press.

___ *Seeing Seattle* (1994b) Seattle: University of Washington Press.

Salinger, J. D. (1958) *The Catcher in the Rye*. London: Penguin.

Sanford, Christopher (1996) *Kurt Cobain*. London: Vista.

Santiago-Lucerna, Javier (1999) '"Frances Farmer will have her revenge on Seattle": Pan-capitalism and Alternative Rock', in J. S. Epstein (ed.) *Youth Culture: Identity in a Postmodern World*. Oxford: Blackwell, 189–90.

Santos, Noelia (1996) 'The Trends Start Here', *Independent*, 9 November.

Sassen, Saskia (1991) *The Global City*. New Jersey: Princeton University Press.

Savage, Dan (1996) 'I Love Seattle Commons', *Stranger*, 1 May, 6.

Savageau, David and Geoffrey Loftus (1997) *Places Rated Almanac*. New York: Macmillan Travel.

Scardapane, Dario (1992) 'rock auteur', *Vogue*, September, 330–2.

Schalit, Joel and John Brady (1993) 'Seattle Uber Alles', *Bad Subjects*, 9, November. www.badsubjects.com.

Schiller, Lawrence, Don Cravens and Ralph Crane (1962) 'Startling world of Century 21: Seattle soars off into space', *Life*, 4 May 1962, 1–2.

Schmich, Mary T. (1990) 'Now that its secret is out, Seattle is paying the price', *Chicago Tribune*, 29 March, B12.

Schwantes, Carlos Arnaldo (1996) *The Pacific Northwest*. Lincoln: University of Nebraska Press.

Seattle Chamber of Commerce (1997) 'Technology Corridor', 19 October. www.sscchamber.org./

Area/TC.html.

Seattle Magazine (2000) 'Editorial Guidelines: target reader', 4 January. www.seattleinsider.com/partners/seattlemag/editguide.html.

Seltzer, Mark (1998) *Serial Killers*. New York: Routledge.

Sennett, Richard (ed.) (1969) *Classic Essays on the Culture of Cities*. New York, Appleton-Century-Crofts.

Seven, Richard (1989) '"Most Livable City" – Seattle hits top of the metropolitan ratings', *Seattle Times*, 25 October, D1.

Shaefer, David (1987) 'Park Places? Compared with other urban cores, Seattle's downtown doesn't stack up', *Seattle Times*, 5 July, Pacific Section, 8.

Shank, Barry (1994) *Dissonant Identities: The Rock 'n' Roll Scene in Austin, Texas*. Connecticut: Wesleyan University Press.

Shapiro, Joseph P. (1993) 'Just Fix It!', *US News and World Report*, 22 February, 50–6.

Shields, Rob (ed.) (1992) *Lifestyle Shopping*. London: Routledge.

___ (1992) *Places on the Margin*. London: Routledge.

Skelton, Tracey and Gill Valentine (1998) *Cool Places*. London: Routledge.

Skolnick, Arlene (1991) *Embattled Paradise: The American Family in an Age of Uncertainly*. New York: Basic Books.

Slavik, Laura (1999) 'So close, yet it feels so far', *Seattle*, July/August, 27.

Slotkin, Richard (1992) *Gunfighter Nation*. New York: Atheneum.

Smith, Carlton (1991) 'Black babies dying, officials told, Southeast Seattle residents demand action, not more talk', *Seattle Times*, 3 October, B3.

Smith, Henry Nash (1957) *Virgin Land: The American West as Symbol and Myth*. New York: Vintage Books.

Smith, Michael D. (1996) 'The Empire Filters Back: Consumption, Production, and the Politics of Starbucks Coffee', *Urban Geography*, 17, 6, 502–24.

Smith, Michael P. (1992) *After Modernism*. New Brunswick, NJ: Transaction.

___ (1996) *The New Urban Frontier*. London: Routledge.

Smith, Neil and Peter Williams (eds) (1986) *Gentrification of the City*. Winchester, MA: Allen and Unwin.

Soja, Edward (1989) *Postmodern Geographies*. London: Verso.

Solomon, Christopher (1996) 'REI gears up', *Seattle Times*, 24 March 1996.

Stadler, Matthew (1999) 'Delirious Seattle', *Stranger*, 30 July, 19.

Stasio, Marilyn (1998) 'Gumshoes in the Great Outdoors', *New York Times*. 23 August, 27.

Steinbrueck, Victor (1973) *Seattle Cityscape #2*. Seattle: University of Washington Press.

Straw, Will (1991) 'Systems of Articulation, Logics of Change: Communities and Scenes in Popular Music', *Cultural Studies*, 5, 3, October, 368–89.

Stripling, Sherry (1993) 'A State of Bean: Brew Mornings – Seattle's Coffee Culture Exerts Quite a Pull on Java Junkies', *Seattle Times*, 16 May, L1.

Sullivan, Andrew (1993) 'Go West', *New Republic*, 18 October, 50.

Sutton, Jessop (1999) 'The Seattle Coffee Shop in Constantia Village', 7 October. www.human-beams.com/lavie/travel/constantia_coffee.shtml.

Tagg, John (1993) *The Burden of Representation: Essays on Photographs and Histories*. Minnesota: University of Minnesota Press.

Takaki, Ronald (1989) *Strangers from a Different Shore*. London: Penguin.

Takami, David A. (1998) *Divided Destiny: A History of Japanese Americans in Seattle*. Seattle: University of Washington Press.

Tate, Stephanie (1995) 'Starbucks Agrees to Code of Conduct', *Stranger*, 15 March, 9

Taylor, Chuck (1997) 'Seattle's a bit player', *Seattle Times*, 27 April, E12.

Taylor, Quintard (1994) *The Forging of a Black Community*. Seattle: University of Washington Press.

___ (1998) *In Search of the Racial Frontier*. New York: W. W. Norton.

Terry, Jennifer and Melodie Calvert (eds) (1997) *Processed Lives*. London: Routledge.

Thompson, Dave (1994) *Never Fade Away*. New York: St. Martin's Paperbacks.

Toffler, Alvin (1970) *Future Shock*. London: Pan.

___ (1980) *The Third Wave*. New York: Bantam Books.

Tomlinson, John (1999) *Globalisation and Culture*. Cambridge: Polity Press.

Toop, David (1991) *Rap Attack 2*. London: Pluto Press.

Touraine, Alain (1969) *La Societe post-industrially*. Paris: Denoel.

True, Everett (1989) 'Sub Pop, Sub Normal, Subversion: Mudhoney', *Melody Maker*, 11 March, 28–9.

Trumbull, Mark (1994) 'Seattle fights urban sprawl with "villages"', *Christian Science Monitor*, 2 May.

Tucker, Amy and Stephen Tanzer (1995) 'All of the Best and None of the Worst of Seattle', *Forbes*, 8 May, 141–50.

Unterberger, Richie (1998) *Seattle: Mini Rough Guide*. London: Rough Guides.

Updike, Robin (1987) 'Brewing up a marketing plan: purchase of Starbucks was Howard Schultz's idea of a good blend', *Seattle Times*, 16 June, C1.

Urry, John (1995) *Consuming Places*. London: Routledge.

'US Census Data, 1990 Database' (1998) 18 May. www.venus.census.gov/cdrom/lookup.

Klinger-Vartabedian, Laurel and Robert A. Vartabedian (1992) 'Media and Discourse in the Twentieth-Century Coffeehouse Movement', *Journal of Popular Culture*, 26, 3, 211–18.

Verhovek, Sam Howe (1999) 'Seattle is Stung, Angry and Embarrassed as Opportunity Turns to Chaos', *New York Times*, 2 December, 1.

Visser, Margaret (1993) *The Rituals of Dinner: The Origins, Evolution, Eccentricities, and Meaning of Table Manners*. Harmondsworth: Penguin.

Wagner, Stephen (1995) 'Cities that Satisfy', *American Demographics*, September, 18–20.

Walker, Martin (1992) 'Washington State replaces California as haven for infrastructure investment', *Guardian*, 14 September.

Warren, Stacy (1992) 'Disneyfication of the Metropolis: Popular Resistance in Seattle', *Journal of Urban Affairs*, 16, 2, 89–107.

___ (1993) 'This Heaven Gives Me Migraines: The problems and promise of landscapes of leisure', in James Duncan and David Ley (eds) *Place/Culture/Representation*. London Routledge, 173–86.

Watson, Emmett (1989a) 'Mount the Ramparts! Fight Californication!', *Seattle Times*, 30 July, B1.

___ (1989b) 'Most Livable City? It's Time to Berate the Ratings', *Seattle Times*, 26 October, C1.

Wernick, Andrew (1991) *Promotional Culture*. London: Sage.

West, Cornell (1993) *Race Matters*. Boston: Beacon Press.

Westwood, Sallie and John Williams (eds) (1997) *Imagining Cities*. London: Routledge.

White, Richard and Patricia Nelson Limerick (1994) *The Frontier in American Culture*. Berkeley and Los Angeles: University of California Press.

Whittell, Giles (1997) 'The language of hyper-choice has reached its frothy zenith in the coffee culture of America', *The Times*, 20 February, Magazine section. 12.

___ (1998) 'Jobless in Seattle as slump bites', *The Times*, 5 December, 18.

Wilcock, John (ed.) (1997) *Seattle*. Singapore: APA Publications.

Williamson, Don (1990) 'Welcome to Emerald City', *Black Enterprise*, June, 232–8.

Wilson, Alexander (1992) *The Culture of Nature*. Cambridge, MA and Oxford: Blackwell.

Wilson, Elizabeth (1991) *The Sphinx in the City: Urban Life, the Control of Disorder and Women*. London: Virago.

Wrigley, Neil and Michelle Lowe (eds) (1996) *Retailing, Consumption and Capital: Towards The New Retail Geography*. London: Longman.

Wrobel, David M. (1993) *The End of American Exceptionalism: Frontier Anxiety from the Old West to the New Deal*. Lawrence: University Press of Kansas.

___ and Michael C. Steiner (eds) (1997) *Many Wests*. Lawrence: University Press of Kansas.

Young, Stanley (1997) 'Beginnings', in John Wilcock (ed.) *Seattle*. Singapore, APA Publications, 41.

Zinn, Laura (1992) 'Move Over Boomers', *Business Week*, 14 December, 74–82.

Zukin, Sharon (1989) *Loft Living*. New Brunswick, New Jersey: Rutgers University Press.

___ (1990) 'Socio-Spatial Prototypes of a New Organisation of Consumption: The Role of Real Cultural Capital', *Sociology*, 24, 1, February, 37–56.

___ (1992) *Landscapes of Power: From Detroit to Disneyland*. Berkeley: University of California Press.

___ (1995) *The Culture of Cities*. Oxford: Blackwell.

INDEX

UNIVERSITY COLLEGE WINCHESTER
LIBRARY